The Sixth Extinction You Can Stop

The Sixth Extinction You Can Stop

Matthew Petchinsky

1

The Sixth Extinction You Can Stop

By: Matthew Petchinsky

Disclaimer

This book, *The Sixth Extinction You Can Stop*, is provided for informational, educational, and awareness purposes only. It presents complex scientific, environmental, and ecological topics in simplified models and accessible language to ensure broad readability and understanding. While every effort has been made to accurately reflect current research and consensus in the fields of ecology, climate science, conservation biology, and environmental policy, the material is necessarily condensed, interpreted, and simplified for clarity. Simplification may omit nuance, evolving debates, or emerging research.

This book does not constitute professional, scientific, legal, medical, agricultural, environmental engineering, financial, or governmental advice. Readers are encouraged to consult qualified experts, peer-reviewed research, official agencies, and licensed professionals before making significant personal, financial, agricultural, environmental, or policy decisions based on the information presented here. The author and publisher assume no responsibility for actions taken by readers as a result of applying concepts, frameworks, or strategies discussed in this text.

Environmental systems are complex, dynamic, and influenced by regional, cultural, economic, and political variables. Solutions that are effective in one location may not be appropriate in another. Local laws, regulations, and environmental policies vary widely. Readers are solely responsible for ensuring compliance with applicable laws, codes, ordinances, and safety guidelines in their jurisdiction when implementing sustainability, conservation, land management, wildlife interaction, energy use, or waste reduction practices.

References to organizations, public figures, scientific models, technologies, policies, or case studies are included for educational illustration only and do not imply endorsement, affiliation, or approval by those entities. Any statistics or projections cited reflect information available at the time of writing and may change as new research emerges.

The title and thematic framing of "The Sixth Extinction" refers to widely discussed scientific concerns about biodiversity loss and ecosystem destabilization. The phrase is used in an educational context and does not claim that the reader alone can reverse global ecological systems. Rather, the subtitle "You Can Stop" is motivational language intended to emphasize individual and collective agency within broader systemic change.

This book promotes responsible stewardship, ethical environmental engagement, and peaceful civic participation. It does not endorse unlawful activity, vandalism, sabotage, harassment, or any form of environmental extremism. All advocacy encouraged herein is intended to occur within legal, ethical, and constructive frameworks.

Some readers may find environmental discussions emotionally challenging due to themes of climate change, biodiversity loss, habitat destruction, or species decline. The intent of this work is not to induce fear or despair but to cultivate informed empowerment, strategic thinking, and actionable hope.

No guarantees are made regarding ecological outcomes, financial savings, policy changes, biodiversity restoration, carbon reduction, or measurable environmental impact resulting from the application of concepts presented. Outcomes depend on numerous external variables beyond the control of the author or publisher.

By reading this book, you acknowledge that you are responsible for your own decisions, interpretations, and actions. The author and publisher disclaim liability for any direct, indirect, incidental, consequential, or perceived damages arising from the use or misuse of the information contained within this work.

This book is intended to inspire informed awareness, critical thinking, and constructive participation in environmental stewardship—nothing more, nothing less.

Introduction

Extinction isn't abstract. It's happening quietly.

Not in dramatic, cinematic collapse. Not in a single catastrophic moment that splits the sky. It happens in silence—when a familiar bird call no longer echoes at dawn, when insects that once coated windshields no longer appear, when a forest edge thins year after year until it no longer resembles the place it once was. Extinction is not only the disappearance of species recorded in scientific journals; it is the gradual erasure of relationships—between soil and root, pollinator and flower, predator and prey, river and floodplain.

Most people do not wake up thinking about biodiversity. They wake up thinking about bills, family, deadlines, survival. That is understandable. Modern life pulls attention toward the immediate and urgent. Yet biodiversity—the vast web of living organisms interacting across ecosystems—is not a distant environmental issue reserved for scientists and activists. It is the foundation beneath food systems, water security, disease regulation, climate stability, and economic resilience. It is not a side issue. It is the operating system of the planet.

When scientists discuss what is often called "the sixth extinction," they are referring to an unusually accelerated rate of species loss compared to natural background extinction rates observed in the fossil record. Previous mass extinctions were triggered by asteroid impacts, volcanic upheaval, or dramatic climatic shifts. This time, the primary driver is human activity—land conversion, pollution, overexploitation, invasive species spread, and climate disruption. Yet this is where the story becomes more complex than headlines suggest.

Human activity is also the source of restoration.

The same species capable of destabilizing ecosystems is also capable of rewilding landscapes, restoring wetlands, changing consumption patterns, rebuilding soil, protecting pollinators, reforming supply chains, and reshaping local policy. The narrative is not simply destruction. It is influence. And influence works in both directions.

This book takes a grounded look at biodiversity loss—not through abstract global panic, but through accessible understanding. You will not find exaggerated claims or apocalyptic predictions here. Instead, you will find simplified ecological models that clarify how living systems function, why they destabilize, and where leverage points exist. Complexity does not have to mean incomprehensible. Ecological systems follow patterns. When those patterns are understood, meaningful intervention becomes possible.

The truth is both sobering and empowering: large-scale environmental change emerges from aggregated small-scale decisions. Land use decisions. Dietary patterns. Purchasing habits. Waste streams. Gardening choices. Water management. Local zoning. School programs. Community restoration. Municipal policy. These may seem minor in isolation. They are not minor in accumulation.

There is a misconception that biodiversity collapse is a problem too vast for individuals to influence. That only governments or multinational corporations matter. While systemic reform is undeniably critical, history shows that systemic shifts are often preceded by local momentum. Markets respond to demand. Policy responds to pressure. Culture responds to narrative. And narrative begins at the level of conversation and example.

You cannot single-handedly regulate global fisheries. But you can influence local food systems.

You cannot personally rewrite international climate agreements. But you can reshape your household's ecological footprint.

You cannot restore every forest. But you can protect a corridor, plant native species, reduce chemical runoff, support regenerative practices, and shift community norms.

Ecosystems function through networks. Networks are strengthened node by node. Individuals are nodes. Communities are clusters. Regions are systems of clusters. The scale expands, but the principle remains consistent: small stabilizing actions, when multiplied, alter trajectories.

This book does not promise that you alone will halt global extinction trends. It does assert something more realistic—and more powerful: that measurable impact occurs when informed individuals act within their sphere of influence and coordinate locally. Biodiversity recovery has already occurred in multiple regions around the world when pressure was reduced and habitat restored. Species rebound when given space and time. Rivers heal when pollution decreases. Soil regenerates under changed management. Nature is resilient, but it requires interruption of harmful patterns.

You will explore:

- How biodiversity loss actually unfolds—gradually, systemically, and often invisibly.
- Why ecosystems collapse not from one factor, but from cumulative stress.
- The economic and social structures that accelerate or reduce ecological strain.
- Practical leverage points where individuals and communities create disproportionate positive impact.
- How local restoration contributes to regional and global stabilization.

Throughout this book, simplified frameworks will be used to explain ecological feedback loops, tipping points, and recovery dynamics. These models are not oversimplifications meant to diminish scientific rigor; they are tools to make patterns visible. When patterns are visible, agency increases.

Fear can paralyze. Clarity mobilizes.

Extinction is not an event that will be televised in one final dramatic broadcast. It is a quiet thinning. A reduction in resilience. A weakening of networks. But networks can be strengthened. Soil can be rebuilt. Native species can return. Pollinators can recover. Food systems can shift. Policies can evolve.

The question is not whether biodiversity loss is real. The evidence is overwhelming. The question is whether disengagement is inevitable.

It is not.

This book begins from a simple premise: informed, coordinated, persistent local action matters. Not symbolically. Not performatively. Structurally. Ecological systems respond to pressure changes. When harmful pressure decreases and regenerative pressure increases, trajectories shift.

Extinction is happening quietly.

Restoration can happen quietly too.

Chapter 1: What Extinction Really Means

Extinction Is More Than the Death of a Species

Extinction is often defined in the simplest possible terms: when the last individual of a species dies. Scientifically, that is accurate. But in practice, extinction is not a single event. It is a process—gradual, layered, and often invisible. By the time a species is officially declared extinct, the unraveling has usually been happening for decades, sometimes centuries.

Long before the final individual disappears, populations shrink. Genetic diversity narrows. Habitats fragment. Reproduction declines. Food sources become scarce. The species becomes functionally extinct in certain regions even if a few individuals still survive elsewhere. Extinction begins as weakening before it becomes disappearance.

When we think of extinction as a dramatic, final moment, we miss the early warning signals. We fail to recognize that ecological decline often looks ordinary at first. Fewer birds in the sky. Fewer fish in the water. Fewer insects around lights at night. The absence feels subtle—until it accumulates.

Extinction also represents the collapse of relationships. Every species is embedded within a web of interactions. Pollinators support plants. Plants support herbivores. Herbivores support predators. Soil organisms support nutrient cycles. When one species disappears, its relationships disappear too.

Understanding extinction as a process rather than a singular event reframes responsibility. If extinction unfolds gradually, it can also be interrupted gradually. The earlier the intervention, the greater the chance of stabilization.

The Difference Between Natural Extinction and Accelerated Extinction

Extinction is not new to Earth. Species have always emerged and disappeared. Background extinction—the natural rate at which species vanish due to evolutionary and environmental pressures—has occurred for hundreds of millions of years. Ecosystems adapt, evolve, and reorganize.

Mass extinctions, however, are different. In the fossil record, five major mass extinction events dramatically reduced biodiversity in relatively short geological periods. These events were triggered by catastrophic forces such as asteroid impacts, volcanic eruptions, and rapid climate shifts beyond the adaptive capacity of life at the time.

What distinguishes the current biodiversity crisis is speed. The rate at which species are disappearing today is significantly faster than natural background levels. Habitat destruction, pollution, overfishing, invasive species introduction, and climate disruption are accelerating change beyond what many organisms can adapt to.

Rapid change reduces evolutionary flexibility. Species that might adapt over thousands of years cannot adjust within decades. Migration corridors are blocked. Genetic diversity shrinks before adaptation can occur. Ecosystems lose resilience because the pace of disruption outstrips the pace of recovery.

Recognizing this difference matters. If extinction were purely natural and slow, human influence would be limited. But when extinction is accelerated by human systems, human systems also become the key variable in slowing it down.

Extinction as Ecological Domino Effect

No species exists in isolation. Each one plays roles—some visible, others subtle. Certain species regulate populations. Others cycle nutrients. Some stabilize soil. Some disperse seeds across vast distances. These functions create ecological balance.

When a species disappears, its absence triggers cascading effects. A predator's extinction can lead to prey overpopulation, which may result in vegetation collapse. The loss of pollinators can reduce plant reproduction, affecting entire food webs. Even microorganisms, when diminished, alter soil fertility and water filtration.

These cascading effects are sometimes referred to as trophic cascades or ecological chain reactions. The removal of one link destabilizes the structure. In resilient ecosystems, diversity absorbs shocks. In stressed ecosystems, shocks compound.

Extinction does not always create immediate collapse. Often, systems appear stable for years before tipping points are reached. This delay can create a false sense of security. By the time consequences become visible, restoration becomes more difficult.

Understanding extinction as a domino effect clarifies why biodiversity matters beyond moral concern. It is about system stability. The more diverse an ecosystem, the more buffers it has against disruption.

Functional Extinction: When Species Still Exist but No Longer Matter

A species does not have to be completely gone to be ecologically irrelevant. Functional extinction occurs when population numbers are so low that the species no longer fulfills its ecological role. Even if individuals remain, their impact disappears.

For example, a tree species reduced to scattered individuals may no longer provide sufficient canopy cover to regulate temperature. A pollinator species reduced to minimal numbers may no longer support widespread plant reproduction. A predator reduced below sustainable thresholds may fail to regulate prey populations.

Functional extinction often precedes official extinction. It is a warning stage. The species still exists biologically, but ecologically it has already vanished. Recovery becomes more difficult as genetic diversity narrows and breeding pools shrink.

This concept is critical because it reveals that biodiversity decline is not binary. It is not simply alive or extinct. It exists along a spectrum of diminishing influence. The weakening of ecological roles destabilizes systems long before final disappearance.

When we measure biodiversity loss, we must look beyond species counts. We must examine population health, habitat connectivity, and functional contribution.

Human Systems and Invisible Extinction

Much of modern extinction unfolds out of sight. It happens in forests cleared for agriculture, oceans depleted by industrial fishing, wetlands drained for development, and rivers altered by dams. The impacts are often geographically distant from the consumers driving demand.

Global supply chains obscure cause and effect. A product purchased in one region may contribute to habitat destruction in another. Over time, the cumulative pressure reduces biodiversity in ways that are difficult to trace back to individual choices.

Urban environments can further disconnect people from ecological feedback. When food appears in grocery stores and water flows from taps without visible source, the dependence on functioning ecosystems becomes abstract. This abstraction reduces urgency.

Yet every economic system rests on ecological foundations. Agriculture depends on soil organisms and pollinators. Fisheries depend on breeding grounds and water quality. Pharmaceutical innovation depends on biological diversity. Even climate regulation depends on forests and oceans.

Recognizing the link between daily systems and ecological health restores visibility. Extinction is not separate from economic or social systems—it is entangled within them.

Extinction and the Meaning of Loss

Beyond ecological mechanics lies something deeper: extinction represents irreversible loss. Once a species disappears, millions of years of evolutionary history vanish. Unique genetic adaptations, biochemical compounds, and ecological interactions cannot be recreated.

This loss is not only scientific; it is cultural and ethical. Species are woven into human history—through food systems, mythology, medicine, art, and identity. Their disappearance reshapes landscapes and collective memory.

There is also a psychological dimension. Biodiversity loss contributes to what some researchers call ecological grief—the quiet awareness that something foundational is thinning. Even when unspoken, this awareness can influence emotional well-being.

However, acknowledging loss does not require despair. It requires clarity. Loss becomes motivation when understood properly. Preventing further disappearance becomes an act of stewardship rather than panic.

Extinction reminds us of permanence. But it also clarifies responsibility. What remains is still vast. What can still be protected is significant.

Journal Prompts

1. When was the last time you noticed fewer birds, insects, or animals in a familiar place?
2. What ecosystems do you rely on daily without consciously thinking about them?
3. How does viewing extinction as a process rather than an event change your perspective?
4. What local species are native to your region that you know little about?
5. How might your consumption habits indirectly affect distant ecosystems?
6. What does the idea of functional extinction mean to you personally?
7. Which natural spaces from your childhood have changed over time?
8. How does biodiversity connect to your sense of security or stability?
9. What would ecological stewardship look like at a household level for you?
10. What responsibility do you believe individuals hold within large environmental systems?

Closing Reflection

Extinction is not merely the disappearance of rare animals in distant forests. It is the slow weakening of living networks that support every aspect of human life. It begins quietly, often unnoticed, but its consequences expand outward through ecological relationships.

Understanding what extinction truly means shifts the narrative from abstract catastrophe to systemic change. It reveals early warning signs. It highlights functional decline. It exposes hidden connections between economic systems and ecological health.

When extinction is understood clearly, it becomes less overwhelming and more actionable. Processes can be interrupted. Systems can be stabilized. Roles can be restored.

The first step in prevention is comprehension. What we see clearly, we can influence deliberately.

Chapter 2: The Disappearing Invisible

The Smallest Life Forms Hold the Largest Influence

When people think about extinction, they often picture large animals—elephants, whales, tigers, polar bears. These species are visible, charismatic, emotionally compelling. Yet the majority of life on Earth is small, microscopic, or easily overlooked. Bacteria, fungi, plankton, insects, soil organisms—these are the unseen engineers of planetary stability.

Microorganisms regulate nutrient cycles that make plant life possible. Soil bacteria break down organic matter and convert it into forms plants can absorb. Fungi form mycorrhizal networks that connect tree roots, allowing forests to share nutrients and chemical signals. Without these microscopic partnerships, forests would struggle to survive.

In oceans, phytoplankton produce a significant portion of the oxygen in the atmosphere. These microscopic organisms form the foundation of marine food webs. When plankton populations decline due to warming waters or pollution, the ripple effects move upward through fish populations, seabirds, and marine mammals.

Insects—often dismissed or disliked—play critical roles in pollination, decomposition, and pest regulation. The global decline in insect biomass has profound implications for food systems and ecosystem stability. Their disappearance may not make headlines, but it destabilizes entire ecological structures.

The invisible does not mean insignificant. In ecological systems, the smallest components often hold the greatest structural responsibility. When they weaken, everything above them becomes unstable.

The Silent Decline of Insects and Pollinators

Across multiple regions of the world, studies have documented significant declines in insect populations. This includes bees, butterflies, beetles, moths, and countless other species that form the backbone of terrestrial ecosystems. The reduction is gradual, but consistent.

Pollinators are especially critical. Many crops depend on bees and other insects for successful reproduction. Fruits, vegetables, nuts, and seeds rely on pollination to produce yields. Without healthy pollinator populations, agricultural systems become fragile.

The causes of decline are layered. Habitat loss removes flowering plants and nesting areas. Pesticides disrupt neurological systems and reproduction. Climate change shifts blooming seasons, creating mismatches between pollinators and plants. Urban expansion fragments landscapes that once provided continuity.

Insect decline also affects species higher in the food chain. Birds that rely on insects for feeding their young struggle to find sufficient nutrition. Amphibians and small mammals lose essential food sources. The loss compounds across trophic levels.

Because insects are small and often unnoticed, their decline can progress without widespread public awareness. Yet their disappearance may be one of the most consequential environmental shifts underway.

Soil: The Living Foundation Beneath Our Feet

Soil is not inert dirt. It is a living ecosystem composed of minerals, organic matter, bacteria, fungi, nematodes, arthropods, and countless microscopic organisms. A single teaspoon of healthy soil can contain billions of microorganisms.

These organisms regulate nutrient availability, support plant growth, filter water, and store carbon. Soil biodiversity directly influences crop productivity, erosion control, and climate regulation. Without healthy soil life, agricultural systems degrade rapidly.

Modern agricultural practices—such as monocropping, excessive tillage, and heavy chemical inputs—can disrupt soil ecosystems. Over time, soil loses structure, organic matter declines, and microbial diversity shrinks. Yields may initially remain stable due to synthetic inputs, but long-term resilience weakens.

Erosion further accelerates the problem. Topsoil, which takes centuries to form, can be lost in a matter of years under unsustainable practices. Once soil biodiversity declines, recovery becomes slower and more resource-intensive.

Protecting soil biodiversity is not merely an environmental goal; it is a food security strategy. Healthy soil underpins both ecological and economic stability.

Ocean Life: Decline Beneath the Surface

The oceans cover more than two-thirds of the planet, yet much of their biodiversity remains unseen. Coral reefs, plankton communities, deep-sea ecosystems, and coastal wetlands operate largely beyond everyday observation.

Coral reefs, often called the "rainforests of the sea," support immense biodiversity. Rising ocean temperatures and acidification stress coral systems, leading to bleaching events. When reefs decline, the fish and invertebrate species that depend on them lose habitat and breeding grounds.

Overfishing removes key species from marine food webs. When predator populations collapse, prey species may overpopulate, altering ecological balance. Industrial fishing practices can also damage seafloor habitats that take decades to recover.

Plastic pollution and chemical runoff introduce additional stressors. Microplastics enter marine food chains, affecting organisms at multiple levels. Nutrient runoff from agriculture can create dead zones—areas depleted of oxygen where few species survive.

Because much of ocean decline occurs below the surface, its urgency can feel distant. Yet marine systems regulate climate, support fisheries, and contribute significantly to oxygen production. Invisible does not mean disconnected.

Microbial Diversity and Human Health

The disappearing invisible is not limited to wild ecosystems. Microbial diversity also plays a role in human health. The human microbiome—composed of trillions of microorganisms living on and within the body—interacts with immune systems, digestion, and even neurological processes.

Environmental biodiversity influences human microbial exposure. Natural environments introduce diverse microorganisms that help calibrate immune systems. Reduced exposure to biodiverse environments may correlate with increased rates of allergies and autoimmune conditions.

Urbanization and sanitized environments, while beneficial in many ways, can reduce microbial diversity exposure. Antibiotic overuse further alters microbial balance, both environmentally and within the body.

The link between ecological health and human health underscores interconnectedness. When biodiversity declines outside, internal ecosystems may also shift. Health systems are nested within ecological systems.

Understanding this connection reframes biodiversity as a public health issue, not solely a conservation concern. Invisible organisms influence visible outcomes.

Why the Invisible Disappears First

Small organisms often reproduce quickly, but they are also highly sensitive to environmental change. Because of their size and ecological specialization, they respond rapidly to stressors such as pollution, temperature shifts, and habitat fragmentation.

In many cases, invisible species decline before larger organisms do. They function as early warning indicators. When insect populations drop or soil microbial diversity decreases, it signals systemic imbalance.

Unfortunately, because these organisms are less visible and less charismatic, they receive less protection. Conservation efforts historically focused on large mammals and birds, leaving foundational species under-monitored.

Data gaps further complicate understanding. Many microbial and invertebrate species remain undocumented. Extinction may occur before species are even formally identified.

Recognizing the vulnerability of the invisible shifts conservation priorities. Protecting biodiversity requires attention to entire systems, not just flagship species.

Journal Prompts

1. Which invisible or small organisms have you rarely considered but rely on daily?
2. How often do you think about soil as a living ecosystem?
3. What changes have you noticed in insect populations in your area?
4. How might local pesticide use affect invisible biodiversity?
5. In what ways does ocean health influence your daily life, even indirectly?
6. What role does microbial diversity play in your understanding of health?
7. How might urban design support or reduce biodiversity?
8. Why do you think invisible species receive less public attention?
9. What local actions could protect pollinators in your community?
10. How does knowing about invisible decline change your sense of urgency?

Closing Reflection

The most powerful systems on Earth often operate quietly. Soil organisms build fertility beneath our feet. Plankton generate oxygen far from view. Insects sustain food systems without recognition. Microbes regulate cycles that make life possible.

When these invisible systems weaken, consequences surface slowly but profoundly. Crops falter. Fisheries decline. Health shifts. Ecosystems destabilize. Because the decline is subtle, response can be delayed.

To protect biodiversity, attention must expand beyond what is easily seen. The foundation of life is intricate and interdependent. Safeguarding the invisible safeguards everything built upon it.

Awareness is the first restoration tool. What we choose to see determines what we choose to protect.

Chapter 3: Keystone Species Collapse

What Makes a Species "Keystone"

Not all species influence ecosystems equally. Some organisms play disproportionately large roles relative to their population size. These are known as keystone species—species whose presence shapes entire ecological communities.

The term "keystone" comes from architecture. In an arch, the keystone is the central stone that locks all other stones into position. Remove it, and the structure weakens or collapses. In ecosystems, a keystone species performs a similar stabilizing function.

Keystone species may be predators, herbivores, pollinators, engineers, or even plants. Their influence is not about dominance in number but impact in function. Their activities regulate population sizes, maintain habitat structure, or facilitate nutrient distribution.

When keystone species decline, ecosystems reorganize rapidly. Secondary species may surge or crash. Vegetation patterns may change. Water systems may shift. The absence creates imbalance that radiates outward.

Understanding keystone species shifts the conversation from saving everything equally to protecting stabilizers first. When stabilizers remain intact, ecosystems retain resilience.

Predators as Regulators of Balance

Apex predators often serve as keystone species because they regulate prey populations. When predator numbers are healthy, they prevent overgrazing and maintain biodiversity. Their influence cascades through food webs.

Without predators, herbivore populations can expand beyond sustainable limits. Overgrazing reduces plant diversity and damages habitats. Soil erosion may increase. Other species that rely on vegetation for shelter or food may decline.

This dynamic, known as a trophic cascade, illustrates how top-level changes reshape entire ecosystems. Predators indirectly protect plant communities by controlling herbivore behavior and numbers.

Predators also influence prey movement patterns. The mere presence of a predator can alter grazing locations, creating zones of recovery for vegetation. This behavioral regulation stabilizes landscapes.

When predator populations collapse due to hunting, habitat fragmentation, or poisoning, ecosystems often experience delayed but dramatic shifts. Reintroducing predators in some regions has demonstrated how powerful their regulatory role can be.

Ecosystem Engineers: Builders of Habitat

Some keystone species physically shape their environment. These organisms are often called ecosystem engineers. Their activities create or maintain habitats for other species.

Beavers, for example, build dams that create wetlands. These wetlands support fish, amphibians, birds, and plant species that would not thrive in dry conditions. The beaver's activity increases biodiversity rather than reducing it.

Certain large herbivores open grasslands by grazing and trampling vegetation. This prevents forest overgrowth and maintains habitat for species adapted to open spaces. Without these herbivores, ecosystems may transition into entirely different states.

Even small engineers, such as burrowing animals, aerate soil and influence water infiltration. Coral species build reef structures that shelter thousands of marine organisms. The physical structures created by engineers become biodiversity hubs.

When ecosystem engineers disappear, the habitats they maintain often degrade. Wetlands dry, grasslands overgrow, reefs collapse. The loss extends far beyond a single species.

Pollinators and Seed Dispersers as Silent Keystone Forces

Pollinators and seed dispersers often operate quietly but are critical to ecosystem continuity. Many plants rely on specific insects, birds, or mammals to reproduce and spread.

If pollinators decline, flowering plants may fail to produce seeds. Over time, plant diversity decreases. As plant diversity shrinks, herbivores lose food sources, and the chain reaction continues.

Seed dispersers help maintain forest structure. Birds and mammals carry seeds across distances, enabling plant migration and regeneration. Without dispersal, forests may become less diverse and more vulnerable to disease.

These keystone roles are subtle but powerful. Unlike predators, their impact may not appear dramatic. Yet their absence weakens ecological regeneration mechanisms.

Protecting pollinators and seed dispersers protects the long-term reproductive capacity of ecosystems.

The Economics of Keystone Collapse

Keystone species collapse is not only an ecological issue—it has economic consequences. Fisheries decline when predator-prey relationships destabilize. Agriculture suffers when pollinators diminish. Tourism declines when iconic species disappear.

Healthy ecosystems provide services often referred to as ecosystem services: water purification, soil fertility, flood regulation, carbon storage, and pest control. Keystone species help regulate these services.

When keystone collapse occurs, costs rise. Artificial pollination, water filtration systems, soil amendments, and pest control measures become necessary replacements. These substitutes are often expensive and less effective.

Economic systems frequently undervalue keystone species because their contributions are indirect. Yet their loss reveals hidden dependencies.

Recognizing economic interdependence with keystone species reframes conservation as an investment in long-term stability rather than a charitable act.

Preventing Collapse Through Early Intervention

Keystone collapse rarely happens overnight. Warning signs often appear years before full ecosystem destabilization. Population declines, habitat fragmentation, and reproductive failures serve as indicators.

Monitoring keystone populations allows for earlier intervention. Conservation strategies may include habitat restoration, hunting regulation, pollution reduction, or species reintroduction.

Community-level efforts can also play a role. Protecting local habitats, reducing pesticide use, supporting conservation organizations, and advocating for sustainable land policies all contribute to keystone stability.

Restoration examples from various regions demonstrate that ecosystems can recover when keystone species are supported. Recovery may take time, but the return of balance is measurable.

The key insight is this: protecting stabilizers stabilizes systems. Collapse is not inevitable when early signals are recognized and acted upon.

Journal Prompts

1. What species in your region might function as keystone species?
2. How would your local ecosystem change if its primary predator disappeared?
3. What ecosystem engineers operate in your area, even if small?
4. How do pollinators influence the food you eat each day?
5. What economic systems near you depend on biodiversity stability?
6. Have you observed habitat changes that might signal keystone decline?
7. What policies could strengthen local keystone species protection?
8. How does understanding trophic cascades change your view of predators?
9. What role could community action play in stabilizing key species?
10. Why do you think some keystone species are misunderstood or feared?

Closing Reflection

Keystone species hold ecosystems together in ways that are not always obvious. Their influence may be indirect, behavioral, structural, or reproductive—but it is foundational.

When these species collapse, the consequences extend outward in cascading effects. Biodiversity declines accelerate. Economic systems strain. Ecological resilience weakens.

Protecting keystone species is one of the most strategic forms of conservation. It focuses on stabilizing core structures rather than reacting to collapse after it spreads.

Understanding their role empowers early action. Stabilize the keystone, and the arch remains standing.

Chapter 4: Local Actions, Global Ripples

The Myth of Individual Insignificance

One of the most persistent beliefs surrounding environmental collapse is that individual action does not matter. The scale of biodiversity loss feels global, systemic, and politically complex. Against that backdrop, recycling a bottle or planting a tree can feel trivial.

This belief is understandable—but incomplete. Ecological systems are built from cumulative inputs. Just as environmental degradation results from billions of small, repeated actions, environmental restoration can emerge from distributed corrective actions.

Markets respond to demand patterns. Policy shifts often follow cultural shifts. Cultural shifts begin with individual behavior and conversation. The idea that individuals are powerless overlooks how social momentum forms.

History demonstrates that systemic change rarely begins as a centralized mandate. It begins as distributed adoption—new habits, new norms, new expectations. Once these reach a tipping point, institutions adjust.

Individual action is not about solving everything alone. It is about participating in momentum. Momentum scales.

Household Ecology: The First Sphere of Influence

The household is the most immediate ecosystem most people control. Energy use, food consumption, waste generation, water practices, landscaping choices—all influence environmental pressure.

Dietary patterns affect land use and biodiversity. Food waste contributes to methane emissions. Household chemical choices influence local waterways. Small adjustments in these areas reduce cumulative impact.

Landscaping choices are especially powerful. Planting native species supports pollinators. Reducing lawn monoculture increases habitat diversity. Avoiding pesticides protects insects and soil organisms.

Water conservation reduces strain on freshwater ecosystems. Rain collection, efficient fixtures, and mindful irrigation reduce runoff and watershed stress.

When households shift practices, they reduce demand for extractive systems and increase demand for regenerative systems. Aggregated across communities, these shifts become measurable.

Community Networks as Multipliers

Local action gains strength when coordinated. Community gardens, local conservation groups, neighborhood cleanups, and school environmental programs amplify individual efforts.

Communities can influence zoning decisions, green space protection, and local agricultural practices. When residents advocate for biodiversity-friendly policies, local governments respond.

Shared knowledge increases participation. Workshops, seed exchanges, compost programs, and habitat restoration days turn abstract concern into collective engagement.

Local businesses also respond to community values. If consumers prioritize sustainable sourcing and biodiversity protection, supply chains adjust.

Community networks transform isolated actions into coordinated movements. Coordination multiplies effect.

Economic Signals and Supply Chain Pressure

Every purchase is a signal within a larger economic network. When consumers support companies committed to sustainable sourcing, habitat protection, and regenerative practices, they influence corporate behavior.

Businesses monitor demand trends closely. Shifts in purchasing patterns influence product development, sourcing decisions, and investment priorities.

While one consumer does not change a corporation overnight, millions of similar consumers do. Economic signals scale rapidly when aligned.

Divestment from harmful practices and investment in regenerative systems create feedback loops. Companies that ignore sustainability risk losing competitive advantage.

Economic pressure is not symbolic. It is structural. Supply chains adjust to protect profitability.

Local Restoration as Climate and Biodiversity Strategy

Restoring degraded land, wetlands, and urban green spaces has measurable impact. Native plant restoration supports pollinators. Reforestation enhances carbon storage and water regulation.

Urban green corridors improve biodiversity and reduce heat island effects. Community tree planting increases shade, air quality, and habitat connectivity.

Wetland restoration filters water and buffers flooding. These actions reduce infrastructure strain and protect communities from climate extremes.

While global climate agreements are important, local restoration creates tangible resilience. It also demonstrates proof of concept—showing that recovery is possible.

Local restoration projects generate visible results. Visible results reinforce participation.

The Ripple Effect: From Local to Global

Ecological systems are interconnected. Waterways connect regions. Migratory species cross continents. Atmospheric systems distribute emissions globally. Actions in one place influence conditions elsewhere.

When multiple communities adopt biodiversity-supportive practices, regional trends shift. Regional shifts influence national policy discussions. National policy influences international agreements.

Cultural narratives spread through media and social networks. Successful local initiatives inspire replication elsewhere. Replication accelerates impact.

The ripple effect operates socially and ecologically. One stabilized watershed influences downstream systems. One protected corridor supports migratory patterns.

Global outcomes are built from local inputs. The scale may feel overwhelming, but the structure is cumulative.

Journal Prompts

1. What environmental practices currently operate within your household?
2. Which small change could you realistically implement this month?
3. Are there community environmental groups you could participate in or support?
4. How do your purchasing habits influence distant ecosystems?
5. What native plants grow in your region?
6. How could your neighborhood increase biodiversity in shared spaces?
7. What local policy decisions affect environmental stability in your area?
8. How might coordinated action feel different from individual effort?
9. What restoration projects inspire you most?
10. What ripple could your next decision create?

Closing Reflection

Local actions are not isolated gestures. They are inputs into interconnected systems—ecological, economic, and social. While no single individual can reverse global biodiversity decline alone, collective local action can shift trajectories.

The myth of insignificance dissolves when patterns are understood. Distributed actions scale. Coordinated efforts multiply. Economic signals restructure systems.

Global ripples begin in kitchens, gardens, schools, and city councils. They begin where influence is immediate and visible.

Extinction unfolds through accumulation. Restoration does too.

Chapter 5: Becoming a Biodiversity Ally

Redefining What It Means to Be an Ally

To become a biodiversity ally is not to adopt a label—it is to adopt a posture. An ally is someone who stands with, supports, and acts in defense of something beyond themselves. In the context of biodiversity, this means aligning daily choices, long-term goals, and community influence with the health of living systems.

An ally understands interdependence. Human survival is not separate from ecological stability. Food, water, medicine, climate regulation, and economic continuity all depend on functioning ecosystems. Supporting biodiversity is not charity toward nature; it is stewardship of shared survival.

Becoming an ally also requires awareness. It begins with education—understanding local species, regional ecosystems, and the pressures affecting them. Without awareness, action becomes unfocused or symbolic rather than strategic.

An ally moves beyond performative gestures. Posting about environmental concern is not the same as restructuring habits or supporting policy shifts. Authentic allyship is sustained, practical, and adaptive.

Most importantly, becoming an ally is not about perfection. It is about participation. No individual can eliminate all ecological impact, but everyone can reduce harm and increase regeneration within their sphere of influence.

Shifting from Consumer to Steward

Modern economic systems position individuals primarily as consumers. Consumption drives growth. Growth drives extraction. Extraction drives biodiversity loss. Becoming a biodiversity ally involves shifting identity—from consumer to steward.

A steward asks different questions before acquiring something: Where did this come from? What ecosystems were affected? Is there a regenerative alternative? Is this necessary? These questions interrupt automatic demand.

Stewardship also includes longevity. Choosing durable goods, repairing instead of replacing, and reducing waste all decrease pressure on natural systems. Extending product lifecycles reduces resource extraction upstream.

Food systems are a central arena for stewardship. Supporting local farmers practicing soil conservation, reducing food waste, and diversifying diets contribute to ecosystem resilience.

The shift is psychological before it is practical. When identity changes, behavior follows. A steward does not aim for zero impact but for net positive contribution where possible.

Advocacy Without Extremism

Being a biodiversity ally does not require confrontation or radicalism. Constructive advocacy operates within legal and ethical frameworks. It focuses on solutions rather than hostility.

Engaging in public comment periods, voting for environmentally responsible policies, and attending community planning meetings are powerful tools. Local governance decisions often shape land use, zoning, and conservation funding.

Writing to representatives, supporting evidence-based initiatives, and collaborating with diverse stakeholders fosters change without polarization. Productive dialogue expands participation rather than alienating potential allies.

Advocacy also includes supporting organizations conducting habitat restoration, species monitoring, and education. Financial contributions, volunteer hours, or skill-sharing strengthen their impact.

Sustainable progress often emerges from steady, informed engagement rather than dramatic confrontation. Biodiversity protection benefits from broad coalitions.

Building Resilient Local Habitats

One of the most tangible ways to become a biodiversity ally is by strengthening local habitats. Whether living in urban, suburban, or rural environments, individuals can influence habitat quality.

Planting native species supports pollinators and local food webs. Native plants evolved with local wildlife and provide appropriate nutrition and shelter. Even small gardens can function as micro-habitats.

Reducing chemical use protects soil organisms and water systems. Composting builds soil health and reduces landfill waste. Creating bird-friendly spaces reduces collision risks and increases survival rates.

Urban biodiversity corridors—such as tree-lined streets and connected green spaces—support species movement. Advocating for green infrastructure in cities increases habitat continuity.

Habitat resilience begins locally but contributes regionally. Small patches connect to larger networks, especially for migratory species.

Educating the Next Generation

Long-term biodiversity protection depends on cultural continuity. Teaching children about ecosystems, species identification, and environmental responsibility builds generational momentum.

Experiential learning is especially powerful. Gardening, hiking, birdwatching, and soil observation foster direct connection to living systems. Connection often precedes protection.

Schools and community centers can integrate biodiversity awareness into curriculum and programming. Even simple lessons about pollination or water cycles increase literacy.

Intergenerational conversations also matter. Sharing stories about environmental change across decades provides perspective. Awareness of shifts over time encourages proactive thinking.

Education transforms concern into informed engagement. A generation raised with ecological literacy is more likely to prioritize systemic solutions.

Measuring Impact Without Losing Perspective

Biodiversity allyship benefits from measurement, but not obsession. Tracking reduced waste, energy use, or habitat restoration progress provides motivation and accountability.

However, focusing solely on metrics can create discouragement if global trends remain challenging. The purpose of measurement is improvement, not perfection.

Celebrate incremental progress. A single restored garden may support dozens of species. A community compost program may reduce significant waste over time.

Perspective matters. Individual actions are part of a mosaic. The goal is cumulative contribution rather than isolated heroism.

Allyship is sustained through balance—clear-eyed awareness of ecological challenges paired with steady commitment to constructive change.

Journal Prompts

1. What does being a biodiversity ally mean to you personally?
2. How can you shift from consumer thinking to steward thinking in daily life?
3. Which local species would benefit most from habitat support in your area?
4. What policy decisions in your community affect biodiversity?
5. How could you involve children or younger generations in ecological education?
6. What purchasing habit could you change to reduce ecological strain?
7. How can you measure your environmental impact without becoming overwhelmed?
8. What local restoration project could you join or initiate?
9. Who in your community shares similar environmental values?
10. What long-term biodiversity goal would you like to commit to?

Closing Reflection

Becoming a biodiversity ally is less about dramatic transformation and more about steady alignment. It is choosing to participate consciously in the systems that sustain life.

It involves awareness, stewardship, constructive advocacy, habitat support, education, and measured persistence. No single action reverses biodiversity decline—but sustained patterns do.

Allyship scales when practiced collectively. The decision to align with living systems influences households, communities, and markets.

The future of biodiversity is shaped not only by global agreements, but by individuals who choose to stand in support of life where they are.

Appendix A

This appendix provides practical, grounded tools for turning awareness into action. While biodiversity is a global concern, protection and restoration are most effective when approached regionally. Ecological systems differ dramatically across climates, geographies, and cultures. What works in a desert ecosystem may not apply in a temperate forest or coastal wetland.

The following sections outline regional protection strategies, foundational habitat restoration principles, and accessible citizen science pathways that empower individuals and communities to contribute meaningfully to biodiversity stability.

Species Protection Actions by Region

Biodiversity threats differ by ecosystem type. Effective action begins with understanding local ecological pressures and opportunities.

Temperate Forest Regions

Temperate forests are often threatened by fragmentation, invasive species, logging pressures, and suburban expansion. Protecting species in these regions requires preserving contiguous habitat corridors. Wildlife depends on connected landscapes for migration, breeding, and genetic diversity.

Supporting native tree and understory plant restoration increases resilience. Removing invasive plants allows native species to recover and supports insects and birds that evolved alongside them.

Reducing pesticide use protects pollinators and soil organisms. Encouraging mixed-age forest growth strengthens habitat complexity and increases biodiversity.

Community-level forest stewardship programs, including trail maintenance and invasive species monitoring, create sustained protection beyond one-time efforts.

Grasslands and Prairie Ecosystems

Grasslands are among the most endangered ecosystems globally, often converted for agriculture or development. Species protection in these regions centers on preserving native grasses and preventing overgrazing.

Rotational grazing practices can support biodiversity when managed properly. Supporting ranchers who adopt regenerative practices reduces soil degradation and supports native plant diversity.

Fire management is critical in prairie systems. Controlled burns, conducted safely and professionally, help maintain ecosystem balance and prevent woody encroachment.

Protecting pollinator corridors in grasslands strengthens ecosystem health and agricultural productivity simultaneously.

Wetlands and Freshwater Systems

Wetlands filter water, buffer floods, and support diverse species. Draining wetlands for development significantly reduces biodiversity and increases flood risk.

Species protection here involves preventing runoff pollution, reducing fertilizer use, and supporting wetland conservation zoning.

Riparian buffer restoration—planting native vegetation along streams—reduces erosion and protects aquatic habitats.

Reducing plastic waste and supporting wastewater treatment improvements helps protect freshwater biodiversity.

Coastal and Marine Regions

Marine ecosystems face overfishing, warming waters, acidification, and plastic pollution. Supporting sustainable seafood sourcing reduces pressure on vulnerable fish populations.

Participating in beach cleanups reduces debris entering marine food webs. Reducing single-use plastics lowers long-term marine contamination.

Supporting marine protected areas strengthens fish populations and reef recovery. These zones allow ecosystems to rebuild without extraction pressure.

Advocating for responsible coastal development prevents habitat destruction of mangroves, salt marshes, and estuaries.

Arid and Desert Regions

Desert ecosystems are sensitive to disturbance. Off-road vehicle damage, water diversion, and invasive species threaten fragile habitats.

Protecting desert biodiversity includes minimizing land disturbance and supporting water conservation policies.

Native plant restoration helps stabilize soil and prevent erosion. Even small disruptions in arid systems can take decades to recover.

Community education about desert ecology reduces accidental harm to sensitive species.

Habitat Restoration Basics

Restoration does not require large-scale projects to begin. Foundational principles apply at multiple scales—from backyard gardens to community initiatives.

Start With Native Species

Native plants support local insects, birds, and mammals. Because they evolved within specific ecosystems, they require fewer artificial inputs such as fertilizers and pesticides.

Researching regional plant guides ensures appropriate selection. Local nurseries and conservation organizations often provide native plant lists.

Avoiding invasive ornamental plants prevents ecological imbalance. Invasive species can outcompete natives and reduce biodiversity.

Native restoration builds food web stability from the ground up.

Improve Soil Health

Healthy soil underpins habitat success. Composting adds organic matter and supports microbial diversity. Reducing tillage preserves soil structure and beneficial organisms.

Avoid synthetic overuse of fertilizers that disrupt natural nutrient cycles. Encouraging natural decomposition cycles strengthens resilience.

Plant diversity improves soil structure and nutrient availability. Diverse root systems enhance water infiltration and carbon storage.

Soil health restoration is foundational to long-term habitat success.

Create Structural Diversity

Habitat complexity increases species diversity. Incorporating trees, shrubs, groundcover, and open spaces creates varied niches.

Dead wood, leaf litter, and natural debris support insects and fungi. Over-sanitized landscapes reduce microhabitats essential to biodiversity.

Water features, even small ones, support amphibians and pollinators. Bird-friendly structures increase survival rates.

Structural diversity enhances ecosystem resilience.

Reduce Chemical Inputs

Minimizing pesticides and herbicides protects non-target species. Even low doses can disrupt insect populations and soil microorganisms.

Integrated pest management approaches prioritize prevention and natural controls.

Avoiding chemical runoff protects waterways and aquatic life.

Reducing synthetic inputs strengthens natural regulation mechanisms.

Monitor and Adapt

Restoration is iterative. Observing plant growth, species return, and soil condition informs future adjustments.

Patience is necessary. Ecosystems rebuild over seasons and years, not weeks.

Tracking changes helps measure progress and refine strategies.

Adaptation ensures restoration remains responsive and effective.

Citizen Science Resources

Citizen science empowers individuals to contribute data to scientific research. Participation improves ecological monitoring and increases public understanding.

Biodiversity Observation Platforms

Mobile apps and online platforms allow individuals to record species sightings. These records contribute to large biodiversity databases used by researchers.

Photographing plants, insects, and animals helps track population trends and migration patterns.

Repeated local observations provide valuable longitudinal data.

Citizen participation strengthens scientific monitoring capacity.

Bird and Pollinator Monitoring Programs

Annual bird counts and pollinator surveys rely on volunteers to collect standardized data.

These programs track population trends and identify areas of concern.

Participation requires minimal equipment and increases species literacy.

Monitoring supports early detection of decline.

Water Quality Testing Initiatives

Community-based water monitoring programs test pH, temperature, and pollutant levels.

Local data informs conservation priorities and municipal decisions.

Participation increases awareness of watershed health.

Water monitoring connects biodiversity to daily life.

Invasive Species Reporting

Many regions maintain reporting systems for invasive species sightings.

Early detection prevents widespread ecological damage.

Reporting contributes to rapid response efforts.

Community vigilance strengthens ecosystem defense.

Educational and Extension Services

Universities and environmental organizations offer workshops, training sessions, and resources for habitat restoration and species identification.

Extension services often provide region-specific guidance.

Learning networks expand practical knowledge.

Education strengthens long-term engagement.

Final Note

Protecting biodiversity does not require waiting for global mandates. It begins with informed regional action, grounded restoration practices, and participation in monitoring efforts.

Each ecosystem has distinct needs. Each community has distinct leverage points. When local protection aligns with scientific understanding, the cumulative effect becomes powerful.

Awareness leads to stewardship. Stewardship leads to resilience.

Message from the Author:

I hope you enjoyed this book, I love astrology and knew there was not a book such as this out on the shelf. I love metaphysical items as well. Please check out my other books:

-Life of Government Benefits

-My life of Hell

-My life with Hydrocephalus

-Red Sky

-World Domination:Woman's rule

-World Domination:Woman's Rule 2: The War

-Life and Banishment of Apophis: book 1

-The Kidney Friendly Diet

-The Ultimate Hemp Cookbook

-Creating a Dispensary(legally)

-Cleanliness throughout life: the importance of showering from childhood to adulthood.

-Strong Roots: The Risks of Overcoddling children

-Hemp Horoscopes: Cosmic Insights and Earthly Healing

- Celestial Hemp Navigating the Zodiac: Through the Green Cosmos

-Astrological Hemp: Aligning The Stars with Earth's Ancient Herb

-The Astrological Guide to Hemp: Stars, Signs, and Sacred Leaves

-Green Growth: Innovative Marketing Strategies for your Hemp Products and Dispensary

-Cosmic Cannabis

-Astrological Munchies

-Henry The Hemp

-Zodiacal Roots: The Astrological Soul Of Hemp

- Green Constellations: Intersection of Hemp and Zodiac

-Hemp in The Houses: An astrological Adventure Through The Cannabis Galaxy

-Galactic Ganja Guide

Heavenly Hemp

Zodiac Leaves

Doctor Who Astrology

Cannastrology

Stellar Satvias and Cosmic Indicas

Celestial Cannabis: A Zodiac Journey

AstroHerbology: The Sky and The Soil: Volume 1

AstroHerbology:Celestial Cannabis:Volume 2

Cosmic Cannabis Cultivation

The Starry Guide to Herbal Harmony: Volume 1

The Starry Guide to Herbal Harmony: Cannabis Universe: Volume 2

Yugioh Astrology: Astrological Guide to Deck, Duels and more

Nightmare Mansion: Echoes of The Abyss

Nightmare Mansion 2: Legacy of Shadows

Nightmare Mansion 3: Shadows of the Forgotten

Nightmare Mansion 4: Echoes of the Damned

The Life and Banishment of Apophis: Book 2

Nightmare Mansion: Halls of Despair

Healing with Herb: Cannabis and Hydrocephalus

Planetary Pot: Aligning with Astrological Herbs: Volume 1

Fast Track to Freedom: 30 Days to Financial Independence Using AI, Assets, and Agile Hustles

Cosmic Hemp Pathways

How to Become Financially Free in 30 Days: 10,000 Paths to Prosperity

Zodiacal Herbage: Astrological Insights: Volume 1

Nightmare Mansion: Whispers in the Walls

The Daleks Invade Atlantis

Henry the hemp and Hydrocephalus

10X The Kidney Friendly Diet

Cannabis Universe: Adult coloring book

Hemp Astrology: The Healing Power of the Stars

Zodiacal Herbage: Astrological Insights: Cannabis Universe: Volume 2

Planetary Pot: Aligning with Astrological Herbs: Cannabis Universes: Volume 2

Doctor Who: Convergence Protocol – The Replicator War

Nightmare Mansion: Curse of the Blood Moon

The Celestial Stoner: A Guide to the Zodiac

Cosmic Pleasures: Sex Toy Astrology for Every Sign

Hydrocephalus Astrology: Navigating the Stars and Healing Waters

Lapis and the Mischievous Chocolate Bar

Celestial Positions: Sexual Astrology for Every Sign

Apophis's Shadow Work Journal: **:** A Journey of Self-Discovery and Healing

Kinky Cosmos: Sexual Kink Astrology for Every Sign

Digital Cosmos: The Astrological Digimon Compendium

Stellar Seeds: The Cosmic Guide to Growing with Astrology

Apophis's Daily Gratitude Journal

Cat Astrology: Feline Mysteries of the Cosmos

The Cosmic Kama Sutra: An Astrological Guide to Sexual Positions

Unleash Your Potential: A Guided Journal Powered by AI Insights

Whispers of the Enchanted Grove

Cosmic Pleasures: An Astrological Guide to Sexual Kinks

369, 12 Manifestation Journal

Whisper of the nocturne journal(blank journal for writing or drawing)

The Boogey Book

Locked In Reflection: A Chastity Journey Through Locktober

Generating Wealth Quickly:How to Generate $100,000 in 24 Hours

Star Magic: Harness the Power of the Universe

The Flatulence Chronicles: A Fart Journal for Self-Discovery

The Doctor and The Death Moth

Seize the Day: A Personal Seizure Tracking Journal

The Ultimate Boogeyman Safari: A Journey into the Boogie World and Beyond

Whispers of Samhain: 1,000 Spells of Love, Luck, and Lunar Magic: Samhain Spell Book

Apophis's guides:Witch's Spellbook Crafting Guide for Halloween

Frost & Flame: The Enchanted Yule Grimoire of 1000 Winter Spells

The Ultimate Boogey Goo Guide & Spooky Activities for Halloween Fun

Harmony of the Scales: A Libra's Spellcraft for Balance and Beauty

The Enchanted Advent: 36 Days of Christmas Wonders

Nightmare Mansion: The Labyrinth of Screams

Harvest of Enchantment: 1,000 Spells of Gratitude, Love, and Fortune for Thanksgiving

The Boogey Chronicles: A Journal of Nightly Encounters and Shadowy Secrets

The 12 Days of Financial Freedom: A Step-by-Step Christmas Countdown to Transform Your Finances

Sigil of the Eternal Spiral Blank Journal

A Christmas Feast: Timeless Recipes for Every Meal

Holiday Stress-Free Solutions: A Survival Guide to Thriving During the Festive Season

Yu-Gi-Oh! Holiday Gifting Mastery: The Ultimate Guide for Fans and Newcomers Alike

Holiday Harmony: A Hydrocephalus Survival Guide for the Festive Season

Celestial Craft: The Witch's Almanac for 2025 – A Cosmic Guide to Manifestations, Moons, and Mystical Events

Doctor Who: The Toymaker's Winter Wonderland

Tulsa King Unveiled: A Thrilling Guide to Stallone's Mafia Masterpiece

Pendulum Craft: A Complete Guide to Crafting and Using Personalized Divination Tools

Nightmare Mansion: Santa's Eternal Eve

Starlight Noel: A Cosmic Journey through Christmas Mysteries

The Dark Architect: Unlocking the Blueprint of Existence

Surviving the Embrace: The Ultimate Guide to Encounters with The Hugging Molly

The Enchanted Codex: Secrets of the Craft for Witches, Wiccans, and Pagans

Harvest of Gratitude: A Complete Thanksgiving Guide

Yuletide Essentials: A Complete Guide to an Authentic and Magical Christmas

Celestial Smokes: A Cosmic Guide to Cigars and Astrology

Living in Balance: A Comprehensive Survival Guide to Thriving with Diabetes Insipidus

Cosmic Symbiosis: The Venom Zodiac Chronicles

The Cursed Paw of Ambition

Cosmic Symbiosis: The Astrological Venom Journal

Celestial Wonders Unfold: A Stargazer's Guide to the Cosmos (2024-2029)

The Ultimate Black Friday Prepper's Guide: Mastering Shopping Strategies and Savings

Cosmic Sales: The Astrological Guide to Black Friday Shopping

Legends of the Corn Mother and Other Harvest Myths

Whispers of the Harvest: The Corn Mother's Journal

The Evergreen Spellbook

The Doctor Meets the Boogeyman

The White Witch of Rose Hall's SpellBook

The Gingerbread Golem's Shadow: A Study in Sweet Darkness

The Gingerbread Golem Codex: An Academic Exploration of Sweet Myths

The Gingerbread Golem Grimoire: Sweet Magicks and Spells for the Festive Witch

The Curse of the Gingerbread Golem

10-minute Christmas Crafts for kids

Christmas Crisis Solutions: The Ultimate Last-Minute Survival Guide

Gingerbread Golem Recipes: Holiday Treats with a Magical Twist

The Infinite Key: Unlocking Mystical Secrets of the Ages

Enchanted Yule: A Wiccan and Pagan Guide to a Magical and Memorable Season

Dinosaurs of Power: Unlocking Ancient Magick

Astro-Dinos: The Cosmic Guide to Prehistoric Wisdom

Gallifrey's Yule Logs: A Festive Doctor Who Cookbook

The Dino Grimoire: Secrets of Prehistoric Magick

The Gift They Never Knew They Needed

The Gingerbread Golem's Culinary Alchemy: Enchanting Recipes for a Sweetly Dark Feast

A Time Lord Christmas: Holiday Adventures with the Doctor

Krampusproofing Your Home: Defensive Strategies for Yule

Silent Frights: A Collection of Christmas Creepypastas to Chill Your Bones

Santa Raptor's Jolly Carnage: A Dino-Claus Christmas Tale

Prehistoric Palettes: A Dino Wicca Coloring Journey
The Christmas Wishkeeper Chronicles
The Starlight Sleigh: A Holiday Journey
Elf Secrets: The True Magic of the North Pole
Candy Cane Conjurations
Cooking with Kids: Recipes Under 20 Minutes
Doctor Who: The TARDIS Confiscation
The Anxiety First Aid Kit: Quick Tools to Calm Your Mind
Frosty Whispers: A Winter's Tale
The Infinite Key: Unlocking the Secrets to Prosperity, Resilience, and Purpose
The Grasping Void: Why You'll Regret This Purchase
Astrology for Busy Bees: Star Signs Simplified
The Instant Focus Formula: Cut Through the Noise
The Secret Language of Colors: Unlocking the Emotional Codes
Sacred Fossil Chronicles: Blank Journal
The Christmas Cottage Miracle
Feeding Frenzy: Graboid-Inspired Recipes
Manifest in Minutes: The Quick Law of Attraction Guide
The Symbiote Chronicles: Doctor Who's Venomous Journey
Think Tiny, Grow Big: The Minimalist Mindset
The Energy Key: Unlocking Limitless Motivation
New Year, New Magic: Manifesting Your Best Year Yet
Unstoppable You: Mastering Confidence in Minutes
Infinite Energy: The Secret to Never Feeling Drained
Lightning Focus: Mastering the Art of Productivity in a Distracted World
Saturnalia Manifestation Magick: A Guide to Unlocking Abundance During the Solstice
Graboids and Garland: The Ultimate Tremors-Themed Christmas Guide
12 Nights of Holiday Magic
The Power of Pause: 60-Second Mindfulness Practices

The Quick Reset: How to Reclaim Your Life After Burnout
The Shadow Eater: A Tale of Despair and Survival
The Micro-Mastery Method: Transform Your Skills in Just Minutes a Day
Reclaiming Time: How to Live More by Doing Less
Chronovore: The Eternal Nexus
The Mind Reset: Unlocking Your Inner Peace in a Chaotic World
Confidence Code: Building Unshakable Self-Belief
Baby the Vampire Terrier
Baby the Vampire Terrier's Christmas Adventure
Celestial Streams: The Content Creator's Astrology Manual
The Wealth Whisperer: Unlocking Abundance with Everyday Actions
The Energy Equation: Maximize Your Output Without Burning Out
The Happiness Algorithm: Science-Backed Steps to Joyful Living
Stress-Free Success: Achieving Goals Without Anxiety
Mindful Wealth: The New Blueprint for Financial Freedom
The Festive Flavors of New Year: A Culinary Celebration
The Master's Gambit: Keys of Eternal Power
Shadowed Secrets: Groundhog Day Mysteries
Beneath the Burrow: Lessons from the Groundhog
Spring's Whispers: The Groundhog's Prediction
The Limitless Mindset: Unlock Your Untapped Potential
The Focus Funnel: How to Cut Through Chaos and Get Results
Bold Moves: Building Courage to Live on Your Terms
The Daily Shift: Simple Practices for Lasting Transformation
The Quarter-Life Reset: Thriving in Your 20s and 30s
The Art of Shadowplay: Building Your Own Personal Myth
The Eternal Loop: Finding Purpose in Repetition
Burrowing Wisdom: Life Lessons from the Groundhog
Shadow Work: A Groundhog Day Perspective
Love in Bloom: 5-Minute Romantic Gestures

The Shadowspell Codex: Secrets of Forbidden Magick
The Burnout Cure: Finding Balance in a Busy World
The Groundhog Prophecy: Unlocking Seasonal Secrets
Nog Tales: The Spirited History of Eggnog
Six More Weeks: Embracing Seasonal Transitions
The Lumivian Chronicles: Fragments of the Fifth Dimension
Money on Your Mind: A Beginner's Guide to Wealth
The Focus Fix: Breaking Through Distraction
January's Spirit Keepers: Mystical Protectors of the Cold
Creativity Unchained: Unlocking Your Wildest Ideas in 2025
Manifestation Mastery: 365 Days to Rewrite Your Reality
The Groundhog's Mirror: Reflecting on Change
The Weeping Angels' Christmas Curse
Burrowed in Time: A Groundhog Day Journey
Heartbeats: Poems to Share with Your Valentine
Dino Wicca: The Sacred Grimoire of Prehistoric Magick
Courage of the Pride: Finding Your Inner Roar
The Lion's Leap: Bold Moves for Big Results
Healthy Hustle: Achieving Without Overworking
Practical Manifesting: Turning Dreams into Reality in 2025
Jurassic Pharaohs: Unlocking the Magick of Ancient Egypt and Dino Wicca
The Happiness Equation: Small Changes for Big Joy
The Confidence Compass: Finding Your Inner Strength
Whispers in the Hollow: Tales of the Forgotten Beasts
Echoes from the Hollow: The Return of Forgotten Beasts
The Hollow Ascendant: The Rise of the Forgotten Beasts
The Relationship Reset: Building Better Connections
Mastering the Morning: How to Win the Day Before 8 AM
The Shadow's Dance: Groundhog Day Symbolism
Cupid's Kitchen: Quick Valentine's Day Recipes
Valentine's Day on a Budget: Love Without Breaking the Bank
Astrocraft: Aligning the Stars in the World of Minecraft

Forecasting Life: Groundhog Day Reflections

Bleeding Hearts: Twisted Tales of Valentine's Terror

Herbal Smoke Revolution: The Ultimate Guide to Nature's Cigarette Alternative

Winter's Wrath: The Complete Survival Blueprint for Extreme Freezes.

The Groundhog's Shadow: A Tale of Seasons

Burrowed Insights: Wisdom from the Groundhog

Sensual Strings: The Art of Erotic Bondage

Whispered Flames: Unlocking the Power of Fire Play

Forgotten Shadows: A Guide to Cryptids Lost to Time

Six Weeks of Secrets: Groundhog Day's Hidden Messages

Shadows and Cycles: Groundhog Day Reflections

The Art of Love Letters: Crafting the Perfect Message

Romantic Getaways at Home: Turning Your Space into Paradise

Purrfect Brews: A Cat Lover's Guide to Coffee and Companionship

The Groundhog's Wisdom: Timeless Lessons for Modern Life

The Shadow Oracle: Groundhog Day as a Predictor

Emerging from the Burrow: A Journey of Renewal

The Language of Love: Learning Your Partner's Love Style

Authorpreneur: The Ultimate Blueprint for Writing, Publishing, and Thriving as an Author

Weathering the Seasons: Groundhog Day Perspectives

Valentine's Day Magic: A Guide to Romantic Rituals

The Shadow Chronicles: Stories of Groundhog Day

Love and Laughter: Fun Games for Valentine's Day

AstroRealty: Unlocking the Stars for Property Success

The Groundhog's Path: A Guide to Seasonal Balance

Groundhog Day Diaries: Reflections in the Shadow

The Groundhog's Light: Illuminating the Path Ahead

Valentine's Traditions from Around the World

AI Wealth Revolution: Unlocking the Trillionaire Mindset

Love Rekindled: Reigniting Passion in Relationships

Single and Thriving: Self-Love on Valentine's Day

Emerald Legends: Mystical Tales of Ireland

Green Alchemy: Harnessing Nature's Magic

The Hearts of Horror: A Valentine's Day Nightmare

The Leprechaun's Guide to Wealth and Wisdom

Dancing with the Sidhe: Celebrating the Otherworld

Shamrocks and Shadows: Mysteries of the Green Isle

Emerald Energy: Harnessing Luck and Growth

The Gingerbread Golem's Valentine: A Sweetheart's Guide to Love and Enchantment

The Celtic Knot: Weaving Life and Destiny

Green Fire: Elemental Magic for St. Patrick's Day

Clover Chronicles: Finding Your Inner Luck

Ireland's Mystical Creatures: A Field Guide

Gingerbread Golem's Love Almanac

Prowl and Thrive: The Lion's Guide to Success

Love Alchemy: Transforming Your Life Through Heart Energy

WORLD DOMINATION: Woman's Rule 3:The New Life

The Midnight Rose: A Guide to Lunar Love Spells

The Forbidden Letters: Writing Your Own Love Prophecy

Luck and Lore: St. Patrick's Day for Modern Mystics

The Green Path: A Pagan Celebration of Renewal

The Dark Architect's Guide to Reprogramming Reality

Prankster's Paradise: A Guide to Harmless Hijinks

Manifest Your Reality: The Law of Attraction Simplified

The TARDIS Owner's Manual: Understanding the Doctor's Ship: *A complete guide to the TARDIS, its technology, secrets, and mysteries*

Starlit Romance: Astrology Secrets for Finding True Love

The Time Lord's Atlas: A Complete Guide to the Whoniverse: *A breakdown of the locations, planets, and dimensions explored in Doctor Who*

Sweetheart Shadows: The Dark Side of Love and Attraction

February Fire: Reigniting Passion in Every Area of Life

The Self-Love Toolkit: 5 Ways to Embrace Who You Are

February Sparks: Ignite Your Dreams in 28 Days

March to Success: A 31-Day Action Blueprint

Ancient Paths: The 13 Sacred Principles of Dino Wicca

Echoes of Tomorrow: Navigating the AI Revolution

The Wellness Blueprint: Balancing Mind, Body, and Soul

Green Horizons: Sustainable Living for a Better Tomorrow

The AI Wealth Code: How to Make Millions with Automation

AI-Powered Creativity: Writing, Art, and Music for Profit

Extinction Rites: Rebirthing Your Soul Through Prehistoric Magick

Sacred Serpents tarot

Celestial Enchantment blank journal

Star Strains

Culinary Journeys: Exploring Global Flavors at Home

The Hollowvale Curse

The Hollowvale Harvest

The Egg of Transformation: Awakening Your Inner Power

Blooming Into Power: A Wiccan Guide to Spring Awakening

The Nightmare Nexus: The Third Doctor's Perilous Haunting

Digital Detox: Reclaiming Your Life in a Connected World

Ostara's Path: Walking the Spiral of Renewal

The Sacred Hare

Financial Freedom: Building Wealth in the Modern Age

Spring's Cauldron: Stirring the Waters of Change

The Hollowvale Pact

Quantum Consciousness: The Science of Reality Shifting

The Hollowvale Hunger

The Sacred Waters Within: A Witch's Guide to Hydrocephalus Magick

The Raven's Nest: Building a Life of Unshakable Stability

AI and the Human Mind: The Future of Intelligence

The Hollowvale Reckoning

Timeless Love: Building and Maintaining Lasting Relationships

The Raven's Roar: Unlocking Unstoppable Confidence

Raven Sight: Awakening Intuition and Inner Wisdom

The Butterfly Effect: Small Changes, Big Transformations

Taming the Boogeyman: How to Conquer Your Inner Fears

The Magick of Green: Awakening Earth's Energy in You

The Entrepreneurial Mindset: Secrets to Business Success

The Hollowvale End

The Shadow Luck Ritual: Reclaiming Power from Your Dark Side

Spring Magick for Beginners: A Simple Guide to Seasonal Energy Work

Doctor Who: The Hollowvale Conundrum

The March of Miracles: Unlocking Synchronicities in Spring

Unveiling the Cosmos: A Guide to Stargazing and Space Exploration

The Ultimate Guide to Surviving an Economic Collapse

The AI Gold Rush: How to Profit from the AI Revolution

Bastet's Shadow: The Hidden Power of Feline Magick

The Bastet Codex: Unlocking the Goddess's Magickal Secrets

Purring Spells: Harnessing Bastet's Healing Frequencies

Bastet's Nine Lives: Rebirth, Transformation, and Immortality Spells

Primal Currents: Hydrocephalus Magick in the Path of Dino Wicca

The Digital Gold Rush: Mastering E-Commerce and Online Sales

Future Shock: Adapting to the Next Decade of Change
The Quantum Mindset: Think Like a Billionaire
Sacred Motherhood: Awakening the Divine Feminine Within
The Mother's Spellbook: Enchantments for Love, Protection, and Prosperity
The Witch's Guide to Parenting: Raising Empowered and Intuitive Children
The Magick of Motherhood: Reclaiming Your Power Through Rituals
The Pagan Path to Self-Love: A Goddess's Guide to Worth and Confidence
Wild Woman Magick: Unleashing Your Primal Power
The Money Magnet Blueprint: Unlocking Unlimited Wealth
Biohacking 101: Unlock Your Body's Full Potential
The Wild Father: A Pagan Guide to Strength and Wisdom
The Sacred Masculine: Unlocking Your Inner Power
The Druid's Compass
The Warrior's Mindset
The Father's Fire
Odin's Path
Ancestral Bonds
The House That Whispers
The Magician's Code
The Wild Hunt
The Green Man's Path
The Altar of Success
The Shadow and the Sword
The High Priestess's Guide to Energy Healing
The Lunar Mother
The Sacred Self-Care Grimoire
The Womb Wisdom Codex
The Wheel of the Mother
The Witch's Guide to Manifestation

The Q2 Reset
The Ultimate Guide to AI-Powered Passive Income
Escape the 9-5
AI Feline Fortunes
The Tear-Stained Grimoire
Razorblade Runes
Cemetery Sirens
The Midnight Wristwatch
The Town That Forgets
AI Horror & Creepypasta
The Hollow Frequency
The Breach Echo
The Quiet Between Worlds
The Sigil of Tharan-Khul
Summon the Vault of Y'ha'ten
The Becoming Codex
The Profit of Az'ra-nar
The Drowned Logos
Echoes of the Eldritch Will
The Deep Ledger
Necronomicon of Networth
Covenant of the Wealthwyrm
The Whisperer's Manifesto
The Rites of Azh-K'luth
The Ark of the Crawling Coin
The Tithe of Shadows
Inkheart Abyss
The Timewinds of Y'ha-nthlei
The Spiral Labyrinth of Azag-Nirrh
The Gallifreyan Heresy of the Black Pharaoh
The Psalms of Nyog-Sotha
Black Rain Alchemy
The Infinite Maw

The Entropic Blueprint
The Oracle of Sh'guul
The Book of Breach
The Drowned Saint's Testament
Dreamcraft of the Sleeper God
The Silence Market
Cthonomics: The Dark Wealth Algorithm
Invocation of the Ten-Eyed King
Wealthbound to the Wyrm Below
Become the Unnameable
Codex of the Sovereign Flame
Rituals of Relentless Becoming
The Shadow Ascends
The Eyes Beneath You
The Will That Wakes Worlds
Silence Is a Weapon
The Mirror That Screams
The Whisper Between Moments
The Mind That Devours Fear
The Myth of the Finished Self
The Architect of Your Madness
The Voice You've Buried
The Discipline of Madness
Stormborn: Awakening Your Inner Tempest
The Mind That Ate Time
Unbind Your Becoming
The Pact You Owe Yourself
The Devourer's Diet
The Acid That Carves the Path
The Tower You Must Burn
The Breath Between Worlds
Speak Like the Deep
The Labyrinth Within

The Spine of the Sea God
Rejection Is a Portal
The Crown You Refused
The Scar Is the Spell
The Lightless Flame
The Habit of Becoming Horrific
ChickenJockey Chaos
The Gatekeeper Within
You Are Not Your Name
The Compass of the Mad
The Archive of Unsent Letters
What the Mirror Can't Show You
The Knife You Needed
Worship Nothing, Become Everything
The Other Voice
The Body the World Forgot
The Vein of the Void
The Black Bone Codex
The Puzzle of the Hidden Self (Millennium Puzzle)
The Eye That Sees the Lie *(Millennium Eye)*
The Ring of Return (Millennium Ring)
The Rod of Relentless Will ***(Millennium Rod)***
The Tally of the Soul (Millennium Tauk/Necklace)
The Key to the Locked Timeline (Millennium Key)
The Scale of Sacred Decisions (Millennium Scales)
Inferno Bites: The UnOfficial Minecraft Lava Cookbook
Rot in the Attic
Prana: The Hidden Force of Your Infinite Self
The Shadow Realm Within: Transforming Darkness Into Destiny
The Borderland Collapse
Claws of Protection: Bastet's Defensive Magick
Mr. Ring-a-Ding's Madness

Yugioh Astrology: Celestial Deckcraft and Duel Destiny (2026–2027 Edition)

The Seal You Signed: Unlocking the Power You Once Feared

The Puzzle of Infinite Minds: Unlocking the Mentalism Hidden Within

The Eye That Mirrors the All: Secrets of Inner Reflection

Doctor Who: The Toymaker's Broadcast

The Rod of Eternal Flow: Commanding the Currents of Vibration

Golden Eyes of Bastet: Enhancing Psychic Vision

The Key of Dual Forces: Balance Within the Polarity

The Scales of Living Rhythm: Timing the Dance of Life

The Necklace of Hidden Cause: Weaving the Webs of Fate

The Ring of Secret Masters: Rising Through the All Within All

Daggers of the Dying Deep

Bastet's Wealth and Fortune Magick: Prosperity Rituals of the Goddess

The Tide that Speaks

The Scrolls of Petosiris: 13 Rituals from the Feathered Eye

Petosiris and the Living Plague of Osirion

Feline Fire: Bastet's Passion and Love Magick

The Star Altar of Petosiris

Bastet's Whiskers: Supernatural Sensory Magick

The Bastet Grimoire

Feeding the Shadows

The Mouthless Prayer

The Spiral Wound

Becoming the Ibis: Lessons from Petosiris's Mind

The Crimson Coven: Cola Magick for Sweet Dominion

Sacred Cat's Paw

The Soda Zodiac: A Flavor for Every Sign

Covenant of the Crawling Flame

The Ink of Ish'Zur
Frothroot: The Thirst That Ate the World
The Golden Glyphs of Prosperity
The Etherbind Codex
Echoes of the Resistance: Reclaiming the You That Survived
Harnessed Minds: Breaking Free from Mental Control
The Eyes in the Smoke
Rootwake: The Carbon Covenant
Skitter Logic: Unlearning the Fear That Built You
Doctor Who: The World That Froths
Rootwake: The Fizz That Rewrites Flesh
Rootwake: Frothfather of the World
The Holly Pact: Blood Beneath the Mistletoe
The 2nd Mass Principle: Building Unbreakable Tribes
Web of Wits: A Survival Guide to Encounters with Anasi the Spider (Aunt Nancy)
The Hexbreaking Handbook: Effective Spells to Remove Curses
Pop Alchemy: Transform Your Life One Sip at a Time
The Mason Code: Leading in Unleadable Times
Petosiris and the Fifth Chamber of Thoth
The Ether Seed Within
The Parent of Tomorrow
Petosiris's Pyramid of Perpetual Wealth
Unlearn the World
Grimoire of the Hollow Tongue
Zodiac Weeds: Finding Your Strain Through the Stars
Aquarius Rises in the Bank
The Sugar God's Smile
The Skinclock Reversal: Biohacking the Face of Time
Debtburn: How to Obliterate What You Owe Forever
The Ice Cream Oracle: What Your Cone Says About Your Future

Silence Is Sovereignty: The Power of Being Unreadable
The Wind That Whispers Through Stone
Path of the Four Directions
Doctor Who: The Maestro's Symphony of Endings
Oxygen Grail: Breathing to Undo the Clock
Zodiacal Collapse: When Stars Devour Time
Teachings from the Red Sand Silence
Doctor Who: Omega – The Broken Equation
Memory Wipe Your Past: Start Over Like a MiB
Mitochondria Prime: Ignite the Core of Youth
A Nest of Roaches
Grub from the Galaxy: MiB Recipes You'll Never Forget
Whiskers of Power: Bastet's Guide to Inner Sovereignty
The Gift Must Cost Them: Negotiation Through Unequal Exchange
The Worm Guys' Wealth Code: Hustle Like an Alien
Bug Out: The Edgar Method for Ruthless Goal Setting
Weaving Life with Spiderwoman's Pattern
Doctor Who: The Rani's Renaissance
How to Get a Job (If You're a Puppygirl)
Speak Through Others: The Art of Proxy Power
T-Rex Your Trauma: Reclaim the Power of Your Primal Self
The Black Briefcase: How MiB Stay Wealthy in Silence
The Thirst Beyond Reason: Understanding Diabetes Insipidus From the Inside Out
Backwoods Witchcraft for Billionaires: The Pagan Path from Dirt Roads to Gold Roads
The Moonshine Grimoire: Fermented Spells for Abundance
The Nameless Geometry
The Wraith's Mixtape: Spellcasting with Songs of Sorrow
The Gallifreyan Gentleman: Silent Wealth, Sonic Confidence, and the Doctor's Code of Living

Sugar Water Confidence

Dino Totems: Spirit Guides from Earth's First Gods

Pumpkin Sigils & Ritual Smoke

The Hózhó Compass

The Distortus Discipline – Order from Inner Mutation

Witchcraft for the Cold-Blooded Soul

Moonlight Over the Blessingway

Water, Interrupted: When the Body Forgets How to Hold On

Deformed but Divine: Embracing Your Asymmetry

Hexes in the Henhouse: The Farm Witch's Guide to Power

The Reversed Flame: A Grimoire of Inverted Power

Salt, Cells, and Sanity: Reclaiming Balance with Diabetes Insipidus

King of the Broken: Crowning the Flawed Self

Moonlight Scars: Emo Moon Magick and the Art of Lunar Sorrow

Noose Spells: Magick of the Hanging Rope

The Goth Cathedral: Rituals of the Broken and Beautiful

The Cretaceous Codex: Dino Spirits and Fossil Magick

The Turkey Bones Ritual

Nine Lives of Manifestation: Unlocking Bastet's Reality Codes

Chaos Muscles: Building Strength from Twisted Pain

Emo Blood Magick: The Veins of Power

The 3AM Flood: Nighttime Urination, Sleep Loss, and the DI Spiral

The Silence Between the Stars

The Lemon Method: Sour Your Way to Success

Starlit Sovereign

The Rite of Twelve Gates: Walking the Gates of the Duat.

Cornbread and Conjure: Recipes for Riches and Rootwork

The Black Rose Rituals: Emo Love Spells and Obsession Magick

The Eternal Goodbye: Death Magick and Rebirth Rituals

Sinkrot: The Drain That Remembers
The Hybrid Mindset: Thinking Beyond Evolution
Bone Dry: Navigating Dehydration Before It Breaks You
Wreathcraft: Winter Sigils and Yuletide Spells
The Rite of Twelve Mouths
S'mores of the Damned: Haunted Recipes for Halloween Night
Rethink the Drink: A Beverage Survival Guide for DI Warriors
The Shadow Chord: Guitar Magick and Sonic
Malformed Yet Majestic: Leading When You Feel Wrong Inside
The Cursed Diary: Ink-Stained Spells of the Emo Soul
Anxiety Off Switch: Emergency Reset for a Fried Nervous System
The Energy Heist: How to Take Back Your Life from Drainers
Beyond the Bathroom: Rebuilding Your Social Life With DI
Coyote's Gift for Decision Makers
Ghoul-Goo: Spooky S'mores You'll Die For
Echoes in the Mirror: Scrying and Self-Destruction Magick
Thread and Thorn: Stitching Spells from Rural Cloth
The Mogwai Mind: Building Untouchable Wealth from Small Ideas
Be Your Own Guide: Trust Your Inner Compass
Hormones in Chaos: The Endocrine Puzzle of Diabetes Insipidus
ChronoCleanse: Detox Your Biological Timeline
The Limb You Hide Is the One That Saves You
Leaves, Bones, and Whisper Spells
The Dinosaur Mindset: Prehistoric Tactics for Modern Wealth
Clawing Back Control: A Bastet Method for Boundaries and Empowerment

Samhain Codes of the Forgotten

Orange Alchemy: The Autumnal Spellbook

Monster Within, Master Without

The Sigil That Sleeps

Hearthcraft: Yule Hearth Sigils and Winter Warmth Spells

Temple of the Purring Moon: Rituals from Bastet's Lineage

When the Earth Remembers Your Footsteps

Food for Mood: Eat to Heal Depression, Not Just Survive It

Witch of the Kudzu Vine: Southern Pagan Wealth Secrets

The Invisible Diagnosis: Advocating for Yourself With DI

The Coffin's Lullaby: Sleep Magick and Nightmares as Portals

Ashes of the Will: Rituals for Erasing the Self

The Fracture Rite: Shattering the Magickal Core

Funeral Candles: The Art of Emo Flame Magick

The Black Grimoire: A Blank Ritual for Writing in Absence

Phantom Thirst: A Halloween Survival Guide for Diabetes Insipidus

The Hollow Crown: Emo Royalty and Magick of the Outcast King

Sandpainting for the Soul

Falling Into Balance: Using the Extra Hour to Recalibrate Life

100 Southern Superstitions That Still Work

The DI Child: A Parent's Guide to Raising Hydrated, Happy Kids

The Jungle of Shame: Escaping the Hidden Lab of Doubt

The Witch's Toast: S'mores Spells and Creepy Confections

The Hoodoo Pantry: Spells from the Mason Jar

The Immortal Fortune: Secret Wealth Lessons from Captain Jack Harkness

The Specter's Embrace: Magick for Channeling Ghostly Lovers

Grateful Waters: A Thanksgiving Survival Guide for Diabetes Insipidus

The Seizure Gospel

Harvestbind: Autumn Sigils and Samhain Spells

Doctor Who: The Algorithmic Awakening

Silent Nights, Steady Waters: A Christmas Survival Guide for Diabetes Insipidus

The Extra Hour Effect: How One Hour Can Transform Everything

Blade Between the Breaths

Running with the Mountain Shadows

Echoes of the Failed Experiment – Becoming a Living Legacy

Swamp Witches & River Ghosts: Real Southern Hauntings

Cornbread Prophecies: Magical Recipes from a Southern Table

Doctor Who: The Signal in the Star

Bloomcraft: Spring Sigils and Equinox Enchantments

Vanisher: How to Disappear from the Internet Without Losing Your Life

The Ultimate Christmas Cookie Cookbook

The Unbreakable Parent: How to Mentally Survive Raising Kids in a Broken World

Time Debt: How to Reclaim the Hours You Didn't Know You Lost

One Month, One Million: Unconventional Cash Creation Tactics

Doctor Who: The Cyber Conversion Protocol

Black Friday Frenzy: The Art of Finding Hidden Online Deals

The Grimoire Buried Under the Apple Tree

Digital Couponing Masterclass: Your Step-by-Step Guide to Black Friday Savings

The Fire Inside the Flint

Shadows on the Ceiling: The Emo Guide to Astral Magick

The Bone-Eaters Beneath Us

Black Friday Bundle Blueprint: Creating Irresistible Product Packages for Maximum Sales
Thirteen Nights of Unbinding
Doctor Who: Engines of Influence
Pumpkin Spice Alchemy: Recipes of Fire and Frost
When Raptor Met Rexy: Prehistoric Lessons in Wild Romance
Oblivion's Candle: Burn the Path Behind You
Campfire Curses and Chocolate: A Halloween S'mores Grimoire
The Wallet That Wasn't Mine: Breaking Free from Inherited Scarcity
The Cola Zodiac: What Your Fizz Says About Your Fate
Frostfire Fortune: Unlocking Winter's Hidden Prosperity
Signal Jamming: Detox Your Brain from Info Overload
The Dust of the Forgotten Rite
The Golden Cat Sigils: Drawing Power in Bastet's Name
Mud Magick: The Prosperity of Getting Dirty
Savory Spicefall: Beyond the Pumpkin Pie
Pumpkin Spice Apothecary Kitchen
The Pumpkin Spice Economy: Turning Trends into Billions
The Ember Veil: 13 Nights of Fire Magick
Financial Freeze: When Earning Triggers Panic
The Gingerbread Alchemist: Sweet Recipes with a Twist
Biosynth Power: Programming Yourself from Scratch
Echoes from the Talking Circle
Truffle Alchemy: Fungus of Fortune
Spores of Power: The Truffle Witch's Handbook
Doctor Who: The Tower That Speaks
The Mycelium Veil: Truffle Magick Unbound
Broke But Ashamed: Healing the Shame Wound of Debt
The Evergreen Mind: Growth Through Winter Stillness
The Black Veil Truffle Rite

The Voice Beneath Symbols
Witchlight on the Bayou: Swamp Spells for Sovereignty
I Grew Up Poor and Now I Sabotage My Success
The Lantern of Hollow Souls
Shutdown Survival: A Citizen's Guide to Thriving When Government Stops
When Abundance Feels Like a Threat: Healing the Hidden Fears of Enough
The Scarecrow Grimoire
Owls of the October Veil
Candlewax and Curses: A Halloween Spellbook
Excavating the Inner Fossil – Transforming Fossilized Thoughtforms into Tools of Power
The Yule Witch's Spellbook: Magick for the Holiday Season
Haunted Christmas: Ghost Stories for Winter Nights
The Christmas Survival Guide: Thriving Through the Chaos
The Bone Orchard Rites
Haunted Commerce: The Ghost in the Marketplace
When Money Triggers the Inner Child
Silent Night with the Silence: A Christmas Thriller
The Lightning Crown: Living and Leading Through Seizures
TARDIS Tidings: Festive Stories from the Whoniverse
Autumn Reset: Slowing Down Without Falling Behind
The Trillion Dollar Soul: Rewriting the Currency of Reality
Truffle Eclipse Magick
Micro-Manifestations: Quick Magick That Works
5-Minute Rituals for Busy Witches
The Marshmallow Massacre: Halloween Treats with a Sinister Twist
Blood and Blackberry Thorns
Magickal Mornings: Rituals for a Blessed Christmas Day
Embers of the All-Hallows Fire
The Pumpkin Market Collapse

The Magick of Leftovers: Transmutation in the Fridge
Frost and Fortune: Christmas Wealthcraft for the Modern Magus
Last-Minute Gifts: Creative Ideas on a Budget
10 Secrets to Surviving Holiday Crowds
The Witch's Dow Jones
The Witches' Graveyard Shift
Pumpkin Sigils: Carving Rites for Midnight Power
The Secret Shopper's Black Friday Guide: How to Get Early Access and VIP Deals
Fever in the Foundation
The Neuralyzer Reset: Reprogram Your Morning Routine
Lightning Lessons of the White Painted Woman
The Enchanted Snowflake
The Magic Sip: Surviving Disney Parks with Diabetes Insipidus
The One Who Draws Through You
Santa's Secrets: A Magical Holiday Planner
Silent Bells, Loud Miracles: Self-Help for the Season of Stillness
The Sugar Spell Codex: Confectionary Magick of Christmas
Autumn Hexcraft: Fallen Leaves, Fallen Barriers
Sleep Like a Fortress: Build a Life Where Insomnia Can't Survive
Nowhere Was Ever Home
Christmas Under the Stars: A Celestial Holiday Guide
Rentproof: How to Never Be Evicted Again
Workless: Escape the 9-5 Without Being Broke
The Gilded Gift: A Wealth Altar for Christmas Morning
The Black Friday Grimoire
The Hollowing: Ceremonies of Spiritual Collapse
Doctor Who: The Nameless Signal
Where the Eagle Hides His Courage

The Dwarf Tyrant's Ascent: Claiming Power in a World of Giants

The Currency of the Dead: Trading with Spectral Bankers

The Icebound Witch: Survival and Magick in the Cold

The Emergency Money Mind: Constant Crisis, Zero Calm

Thirst & Roses: A Valentine's Survival Guide for Diabetes Insipidus

Springing Into Clarity: Mastering the Energy Shift of Daylight Saving

Speed of the Nanotyrant: Agile Strategies for Life's Hunt

The 90-Day Reset: A Science-Backed Guide to Reinvent Yourself Before the New Year

The Coil of Thorns

Timegates Unleashed

Dopamine Fasting 2.0: Rewiring Your Brain in a World of Overstimulation

The Nanotyrannus Mutation: Embracing Your Differentiated Form

Unwritten January: How to Use the Blank Page of the Year to Re-Author Your Life

Feast of the Forgotten Spirits: Ancestor Magick for the Holidays

The Burnout Codex: A System for Thriving Beyond Stress

I Don't Deserve to Have Money

When the King Doesn't See You: The Shadow Predator's Guide

The Reset Ritual :A Guide to Shedding Last Year's Weight—Emotional and Physical—by Spring Cleaning Your Soul

Calendar Alchemy: Learn to Bend Time and Productivity around Your Natural Rhythms

The Mirror Contract: A One-Sentence Agreement with Yourself that Unlocks Real Change

The Nile Formula: Sacred Flow for Millionaire Minds

Forward Momentum: Productivity Rituals for the Lost Hour
Harvest Gold: Wealth Magick of Pumpkin Spice Commerce
The Magick of Christmas Lights: Illuminating Holiday Rituals
Cranberry Crown: Royal Self-Worth through Ritual Meals
Petosiris and the Worms of the Feathered Throne
The Evergreen Equation: Self-Help for Sustained Growth
The Rain Will Wash Us Away: Emo Weather Magick
Snowlight Shadows: Magick Under Frozen Stars
Doctor Who: Merger of Empires
The Light Within the Feast: Magickal Wealth of Yuletide Flame
Momentum Machines: Small Habits Designed to Snowball into Unstoppable Forward Motion
Calm Spells: Magickal Solutions for Anxiety Relief
Santa's Final List: The Naughty Never Leave
The Mystic's Guide to Banishing Worry
The Silence Dividend: Turning Quiet Mornings into Compounding Growth All Year
Charmed Stillness: Rituals to Ease the Mind
Blood Spiral Ascension
The Vanishing Heart Invocation
Crystal Calm: Harnessing Stones for Anxiety
The Nano Tyrant Code: Strength in the Small Predator
The Energy Ledger: Accounting Methods for Vitality, Not Just Money
Snowglobe Mindset: Shake Up Your Thoughts, Settle Into Peace
The Weight of the Feather: Anubis's Codes for Soul Balance
Pocket Peppermint Power: Tiny Rituals for Big Holiday Energy
Shadowkeeper of the Golden Jackal: Anubis's Path to Energetic Protection

The Witch's Breath: Anxiety Relief Through Magickal Breathing

The Tiny Ember: Pocket Motivation for Cold Days

Stocking Sorcery: Mini Spells for Holiday Luck and Prosperity

Warm Hands, Steady Heart: Micro Grounding for Winter Anxiety

The Jackal's Lantern: Guide-Lights for Walking the Dark Path

Anubis's Silent Steps: Rituals for Stealth, Peace & Emotional Quiet

The Bonewake Protocol

Extinction-Proof: How to Outlive Trends, Trauma, and Time

The Fourth Doctor: The Eccentric Wanderer

The Second Doctor: Chaos and Charm in the TARDIS (Patrick Troughton, 1966–1969)

The Third Doctor: Earthbound Defender of Time (Jon Pertwee, 1970–1974)

The First Doctor: Genesis of the Time Lord (William Hartnell, 1963–1966)

The Fourth Doctor: The Eccentric Wanderer ***(Tom Baker, 1974–1981)***

The Fifth Doctor: Elegance and Empathy (Peter Davison, 1982–1984)

The Sixth Doctor: A Kaleidoscope of Conflict (Colin Baker, 1984–1986)

The Seventh Doctor: The Strategist of Shadows (Sylvester McCoy, 1987–1989)

The Eighth Doctor: A Time Lord's Renaissance (Paul McGann, 1996)

The War Doctor: A Time Lord's Burden (John Hurt, 2013)

The Ninth Doctor: The Healer of Worlds
(Christopher Eccleston, 2005)

The Tenth Doctor: The Lonely Hero (David Tennant, 2005–2010)

The Eleventh Doctor: The Madman in a Box (Matt Smith, 2010–2013)

The Twelfth Doctor: The Rebel with a Cause (Peter Capaldi, 2013–2017)

The Thirteenth Doctor: Breaking the Mold

(Jodie Whittaker, 2017–2022)

The Fourteenth Doctor: A New Beginning (David Tennant, 2022–2023)

The Fifteenth Doctor: Rewriting the Stars

(Ncuti Gatwa, 2023–2025)

Santa's Shortcut to Success: Micro Habits for a Better New Year

The Midnight Cocoa Manifesto: Comfort Rituals for Hard Nights

Why I Can't Save: Trauma Blocks to Financial Planning

Dino Brain, Mega Goals: Rewire Yourself for Savage Success.

Jurassic Profit: Unearth Your Inner Millionaire Fossil

Scales & Softness: Dating with Dino Heart

Carnotaurus Dreams.

Red Willow Thinking.

Reindeer Resilience: How to Keep Going When You're Exhausted

Candlewick Courage: Lighting Strength in Dark Seasons.

Elf-Mode Focus: Fast Attention Fixes for Busy Seasons

Futureproof Feelings: Emotional Skill-Building for Resilience No Matter What the New Year Brings

The 90-Day Rewire: Transform Your Neural Pathways Before Spring Blooms

Jolly Rich: Christmas Micro-Money Magick for Fast Prosperity.

Gingerbread Gratitude: Mini Practices for a Happier Holiday

Trauma-Spending: When Money Becomes a Drug

Goal Skeletons: Build Minimalist Frameworks for Resolutions That Last

The Snarl Strategy: Growling Back at a World That Silenced You

The Hourglass Reset: Reclaiming What the Clock Took

Yule Claw: Dino Rites of Light in the Darkness

Sacred Circles: Grounding Rituals for Anxious Souls

Clockstopper's Gut: Rebuilding the Core of Time

Winterfire Drive: Harness the Cold Season's Stillness to Ignite Deep Motivation

The 5-Minute Productivity Hack: Small Steps to Big Results

Light Ascending: A Self-Help Guide to Longer Days

Inner Economy: Trade Self-Doubt for Confidence Like a High-Value Currency

Simple Success: Decluttering Your Life for Better Results

The Resilient Mind: Thriving in Tough Times

The 2026 Playbook: Mastering the Year Ahead

The Spring Sleep Solution: Conquering Fatigue After Time Change.

The Hormonal Compass: Navigate Back to Your Youth

Golden Vision: Eating for Your Eye Color with the Power of Ghee

Lionhearted Love: Strength in Relationships

MiB Mindset: Think Like a Cosmic Agent

The Ladder of Small Wins: Climb Out of Stagnation by Stacking Micro-Victories

The Momentum Hour: Gaining More in Less Time.

The Year of You: A Self-Help Blueprint for Unapologetic Self-Ownership

The Dust That Watches You Back.

The Astral Parent

Shadow of the King: The Nanotyrannus Parable

Tiny Tidings of Triumph: Small Wins That Change Your Life

Dwarf Tyrant Rising: When You're the One They Overlook

The Bank Account Mirror: Healing How You See Yourself Through Money

Cretaceous Black

When Love Costs Too Much: Romantic Triggers and Financial Wounds

Dream-Eater's Guide to Wealth

Unzipped Reality: The Jacket That Opens Portals

The Power of No: How to Set Boundaries and Get Rich

Keeper of the Desert Veil: Anubis's Teachings on Illusion & Truth

Redacted Stargates: Stories the SGC Buried

The Clock That Dripped Time

Doctor Who: The Silent Acquisition

Hairline Manifestation: Unlocking Power at the Edge of Thought

Cash-Eater Crickets: How Insects Predict Economic Collapse

Dead Pet Whisperer: Grief Alchemy from the Other Side

The Love Algorithm Is Broken: Hack Romance Before It Hacks You

The Rite of the 8-Limbed Seal

Melt the Moon: A Fire Ritual to Reshape Emotion

Venomous Miracles: Healing Through the Poison Path

Wealth Without Water: The Post-Aquatic Rich Life

The Butterfly Betrayal: When Change Turns on You

The Quiet Upgrade: How Humans Are Secretly Rebuilding Themselves Without Asking Permission

Nobody Taught Us This: The Missing Human Manual for Money, Energy, and Self-Respect

You Were Never Broken: Why Most People Are Reacting Normally to an Unstable World

After the Noise Ends: How Humans Reset When the World Finally Goes Quiet

The Month That Tells the Truth: Why January Reveals What the Rest of the Year Hides

Tremors Party Unleashed: The Ultimate Guide to a Graboids Birthday Bash

Cold Start Human: Rebooting Motivation Without Forcing Momentum

The Year that doesn't start until they say

Love Without Performance: What Connection Looks Like When No One Is Watching

February Is a Mirror: What Love Exposes About You

How We Learned Love Wrong: Uninstalling Cultural Myths About Connection

You Are Not Too Much: Why Needing Love Is Not a Flaw

The Crooked Fence Covenant: Backwoods Boundaries and Witch Wealth

The Day That Keeps Repeating: Why Your Life Feels Stuck—and How to Shift It Gently

Nothing Changes Until Something Softens: Escaping Repetition Without Violence to Yourself

Same Life, Different Ending: How Tiny Choices Rewrite Repetition

The Loop Isn't Punishment: Why You're Being Asked to Learn, Not Suffer

Panic at the Register: Retail Anxiety and Public Shame

The Secret Formula to Becoming an Influencer in 90 Days

The New Rich: How to Retire by 35

Digital Nomad Secrets: How to Travel the World and Make Millions

Burn Your Waiting Room: Escape the Delay Spiral

Strapped for Pleasure: A Complete Guide to Pegging and Strap-On Play

Locked in Desire: A Guide to Chastity Play and Intimate Control

Burrowing into History: Groundhog Day Through the Ages

The Shadow Forecast: The Science of Groundhog Day

Choice Cascades: How One Decision in January Sets Ripples for the Whole Year

The Anti-Anxiety Blueprint: Mastering Fear and Stress for Success

The Ancestral Path of the Jackal Priest

Winterborn Titans: Giants of Ice and Stone

The Extinction Spire

The Emotion That Made Me Millions

Delete Their Voice: Become Unreachable

The Frostbound Mysteries: Mythical Beings of January's Chill

The Roar Within: Unlocking Your Potential

From Burrow to Sky: Groundhog Day Explained

The Boogey Book: Festive Ward-Off Edition

Fooled Ya! The Psychology Behind Great Pranks

April Fools' Alchemy: Transforming Everyday Items into Pranks

Anchor Habits: Root Yourself in Unshakable Routines That Endure Seasonal Change

Emerald Waters: A St. Patrick's Day Survival Guide for Diabetes Insipidus

The Groundhog's Legacy: Myths, Legends, and Facts

Six Weeks of Mystery: The Truth Behind Groundhog Day

The Upgrade Mind: Tech-Like Patches and Updates for Your Mental Operating System

Moon Blood, Star Flesh: Embracing Your Hybrid Nature

Groundhog Lore and Legend: A Complete Guide

You Are the Lie They Can't Kill

Download a Different Reality

The Quiet Advantage: Winning Without Announcing Yourself
A Year of Gratitude: Unlocking Joy
Touch Grass: A Pagan Reset for a Burned-Out Digital World
Cold Start Wealth: Building Financial Power from a Standing Still Point
Astrology Without Fate: Using Cycles Without Surrendering Control
The Future is Remote: Work Less, Earn More, Live Anywhere
Harvest of Silence
The Algorithm Knows Your Name
April Fools' Day for Introverts
Soft Power Systems: Influence That Doesn't Exhaust You
Practical Magick for Real Life: Focus, Intent, and Follow-Through
January Is a Lie: Designing Momentum Without the Calendar
Universal Flow: Managing DI in High-Energy Theme Parks
Electric Allure: Exploring Erotic Electro-Play
The Masked Embrace: Exploring Sensory Deprivation
The Ultimate April Fools' Survival Guide
Splashproof Survival: Waterparks & Wave Pools with DI
Agile Predator, Still King: Crafting Your Speed in a World That Celebrates Size
The Shadow Feast: Groundhog Day's Sinister Side
The Bug's Business Model
Edge of Bliss: Mastering the Art of Tease and Denial
Midnight Alchemy: Rituals for Transformation Between the Old Year and the New
Punxsutawney Bloodlines: A Groundhog Day Horror
The Burrower: Groundhog Day's Darkest Secrets
Buried Alive: A Groundhog Day Horror Story
Blood Moon Markets: The Witch's Guide to Wealth Magick
The Groundhog's Shadow: What It Means for the World

Seasonal Shadows: The Groundhog Day Phenomenon
The Shadow's Curse: A Groundhog Day Horror Tale
Punxsutawney Revealed: A Guide to Groundhog Day
Safari Hydration: Zoo & Wildlife Park Survival for DI
The Foolproof Prank Planner
Latex Lust: Embracing Fetish Fashion **is a sophisticated**
Whip & Whisper: The Dynamics of Impact Play
The Shadow Burrow Chronicles: Horror Beneath the Ground
Frozen Shadows: A Groundhog Day Thriller
The Groundhog's Revenge: A February Nightmare
Burrow of the Damned
Everyday Enchantment: Magickal Practices for Modern Living
Budget Bliss: The Easy Path to Financial Freedom
Six More Weeks of Terror
The Oxygen Debt
The Ice Veil: Mystical Beasts of January's Shadows
Roots That Remember
Polar Myths: Creatures of the Icy Abyss
The Marginalia Covenant
AI and the Occult: Digital Spirits, Sigils, and Cyber-Witchcraft
Roleplay Realms: Crafting Fantasies and Scenes
The Last Clean River
Lover in the Shadows: A Groundhog Day Tale
Buried Hearts: A Groundhog Day Gothic Romance
Anubis Wept for Petosiris
The Lantern Knows Your Name
Rituals of the Last Shadow
The Bell That Rings for No One
Shadows of Love and Death
The Groundhog's Bride: A Dark Valentine's Tale
Frozen Kisses: A Horror Romance for Groundhog Day
The Creative Spark: Igniting Your Passion Projects

From Cub to King: A Journey of Transformation
Six Weeks to Midnight: A Groundhog Day Romance
Tide of the Unclean
Burrow of Desire: A Groundhog Day Paranormal Romance
The Shadow Magick Almanac: Harnessing Groundhog Day Energies
AI for Emo Influencers: Growing an Online Following Without the Anxiety
AI for the Melancholy Millionaire: Making Money Without Social Interaction
Winter's Grip: A Horror Romance Beneath the Frost
Shadowed Hearts: A Love Buried in Winter's Chill
The Shadow Lover: A Groundhog Day Gothic Love Story
Beneath the Burrow: A Groundhog Day Thriller
The Punxsutawney Conspiracy
The Crimson Pact: Emo Blood Oath Magick
The Shadows of Gallifrey: A Doctor Who Horror Collection
Whispers in the Vortex: Chilling Tales from the TARDIS
AI Ghostwriting for Emo Souls: How to Make Money Writing Without Trying
The Anti-Social Emo's Guide to AI-Powered Freelancing
Emo AI Therapy: Talking to Robots When Humans Suck
Cry & Collect: AI-Powered Passive Income for Emos
The Weeping Angel Chronicles: A Collection of Terror
Dating Without the Mask
The Clockwork Menace: Horror Stories from the Future
Tender Doesn't Mean Weak
Not a Burden, Not a Bonus
Dalek Nightmares: Stories of Ultimate Extermination
The Empty Child's Curse: Tales of Horror and Fear
The TARDIS Ghost Files: Hauntings Across the Universe
Soft Armor, Open Heart
Chosen Gently

Love That Doesn't Rush You

Financial Gaslighting: When You're Made to Feel Crazy About Money.

After the Bees Went Silent

The Soil Knows Your Name

The Quantum Magick Codex: Unlocking the Hidden Laws of Reality

The Gingerbread Golem's Awakening: A Tale of Sweet Revenge

The Gingerbread Golem Chronicles: Guardians of Frost Hollow

The Power of Focus: Eliminate Distractions, Achieve Greatness

Sacred Submission: The Dynamics of D/s Relationships

The Magick of Numbers: Unlocking the Hidden Codes of the Universe

Blood and Bones: The Ancient Magick of Life and Death

Doctor Who: The Collapse Dividend

Thesis of the Unremembered

Desired as You Are

The Energy Budget of Love

Romancing Your Potential: The Love Month Growth Guide

Seen, Not Studied

Screams from the Void: Horror Stories Inspired by Doctor Who

The Silence Is Watching: A Collection of Dread

Under the Floorboards

The House of Feathers and Fangs

Emperor Bones, Hollow Crown

Goreglyphs of the Boiling Deep

The Coven That Screamed Back

The Stacks That Breathe

Love That Adapts

Nightmares on Gallifrey: Frights from the Doctor Who Universe

Apache Ways to Heal Powerfully

Belos Left No Corpses

The Algorithm Hates You (Good.)

Starve the Soul, Feed the Flame

Romance Without Apology

The Carbon Illusion

Your Nervous System is Your GPS

Where Rest Learns Your Name: Six Gentle Bedtime Stories for Minds That Work Differently

Soft Armor for Tender Nights: Six 5-Minute Bedtime Stories for Hearts That Have Been Hurt

Paws of Night, Path of Stars: Six 5-Minute Bedtime Stories Guided by Bastet and Anubis

When the Mind Needs a Lamp: Six 5-Minute Bedtime Stories for Gentle Mental Health Care

Imbolc Roar: Awakening the Dino Flame

The Titan Never Died

Witchflesh Harvest

Become the Thought They Can't Unsee

Golden Threads: The Sensuality of Body Worship

Chains of Desire: Exploring the Beauty of Restraint

Glyphs Written in Organs

Living Below the Line

Palismen Don't Scream—Until They Do

The Boiling Isles Devour the Weak

Hexside Anatomy Lab

Philip Wittebane and the First Body

Money PTSD: When Trauma Hijacks Your Finances

Brother, Bone, and Betrayal

Lipstick Like a Weapon: A Goth Girl's Guide to Emotional Dominance, Boundaries, and Quiet Intimidation

The Scar Is the Portal
Don't Grow—Mutate
Built from Breakdown: The Collapse Code
The Collector's First Lesson
Glyph Failure Syndrome
Philip Wittebane and the Discipline of Lies
The Making of Belos
Flesh Does Not Belong
The Golden Guard Prototypes
Coven Marks as Control Wounds
The Voice Belos Invented
The Boogeyman Within: Shadow Work for Personal Transformation
The Energy Edge: Reclaiming Your Power
Mindful March: A Quick Guide to Intentional Living
Love on Autopilot: Rekindle Your Relationship Instantly
Spring Renewal: Refresh Your Mindset Before the Equinox
Velvet Control: The Subtle Art of Erotic Choking
Erotic Shadows: Playing with Voyeurism and Exhibitionism
The Green Pulse Awakens
Equinox Bones
Immortality Is a Disease
Cold-Blooded Focus
The Invasive Advantage
The Lonely Assassin's Love: A Weeping Angel Romance
Philip Is Still Inside
Become the Signal That Replaces the Noise
When the Soil Says Yes
Shedding What No Longer Protects You
The Shadow Key: Unlocking the Power of Your Dark Self
The Enchanted Chocolate Rituals: Magickal Recipes for Love & Wealth
Love in the Vortex: A Doctor Who Horror Romance

Grimwalker Number Seventeen
The Boy Who Woke Again
Sad Girls Don't Beg: Reclaiming Self-Worth When Melancholy Is Mistaken for Weakness
Raising Pagan Children in a Non-Pagan World
The Pagan Homeschool Exit Strategy
Protecting Childhood While Preparing for Adulthood
Curriculum for the Dead
Too Much for Who? Reclaiming the Power of Intensity
Your Nervous System Is the Algorithm
Memories That Don't Belong to Him
The Golden Guard Template
Time-Locked Hearts: A Gallifreyan Gothic Romance
Ostara's Hatchlings: Fertile Dino Magick
The Dalek's Heart: A Tale of Forbidden Love
The Jackal's Final Game
Where Wild Magic Took My Brother
Where Love Meets Reality
Volume Without Apology
The Mind of Stillness: Petosiris's Hidden Teachings on Power
Doctor Who: The Final Shareholder Meeting
The Archive of Unfinished Lives
AI for Emo Musicians: Automating Your Sad Soundtracks
AI-Generated Emo Art: Selling Melancholy in Digital Form
Ashes Where the Cathedral Slept
The Monarch of Splintered Wings
People-Pleasing Is a Survival Strategy Past Its Expiration Date
Velvet Carnage Covenant
The Intensity Advantage
Attention Is a Finite Resource—You're Overspending It
Emotional Debt Compounds Faster Than Money
The Siege of Thornlit Graves
When the Bone King Wept Black

Comfort Is the Most Expensive Addiction

The AI Gothpreneur: Building a Business Without Any Social Interaction

AI & Tarot: Emo Divination for the Digital Age

Sad AI Aesthetics: Emo Branding & Personal Style

Black Sails Over a Dying Sun

The Crimson Parliament of Wolves

Cathedral of Veins and Thunder

The Orchard of Hanging Saints

The Self-Discovery Path: Find Your True Purpose

You Were Taught to Self-Sabotage Politely

Chrysalis Debt: Reborn into Wealth

Empire of Teeth and Perfume

The Devourer's Bride of Iron Mist

The Dormitory with No Windows

Built Larger Than the Room

The Silence Between Wings

The Manuscript That Grades You

The Librarian Who Refused to Die

Footnotes for a Burning Campus

The Thornblood Accord

The Paranormal Files: True Encounters with Ghosts, Aliens, and the Unexplained

Warfare in the Temple of Shattered Constellations

The Crow's Codex: Unlocking the Secrets of the Black-Winged Messengers

Purring Boundaries: Bastet's Guide to Raising Fierce Children

The Arcane Wealth Formula: Spellcasting for Infinite Abundance

The Celestial Spellbook: Harnessing Planetary Magick for Power and Fortune

Silk Shrouds of the Frostbound Wilds

Windtalker: The Hill Witch's Guide to Whisper Magick
The Doctrine of Psyence: Knowledge Beyond the Veil
The Bone Sovereign's Final War Chant
The Abyssal Altar of Salt and Silver
The Lantern Wilds That Devoured Lovers
The Court of the Sable Scar
Wax & Whispers: The Art of Candle Play
The Billionaire's Blueprint: Unlocking Wealth Through Action
The Cleansing Flow: Unlocking Wellness with Enema Therapy
Seasonal Gratitude: How to Thrive Before Spring
Beyond the Closet Door: Unlocking Your Full Potential by Embracing Fear
The War-Bound Crown
Eternity With Teeth
The Observer Is the Spell: Reality as Psychic Architecture
Mis-Classified Monster: Redefine Yourself
Claws Without Cruelty: The Bastet Discipline Method
Claws Without Cruelty: The Bastet Discipline Method
Temple Mothers: Parenting in the Spirit of the Sacred Cat

www.ingramcontent.com/pod-product-compliance
Lightning Source LLC
LaVergne TN
LVHW022014200926
848481LV00007B/2383

* 9 7 9 8 2 9 5 6 5 9 4 8 5 *